NanoScience and Technology

NanoScience and Technology

Series Editors:
P. Avouris B. Bhushan D. Bimberg K. von Klitzing H. Sakaki R. Wiesendanger

The series NanoScience and Technology is focused on the fascinating nano-world, mesoscopic physics, analysis with atomic resolution, nano and quantum-effect devices, nanomechanics and atomic-scale processes. All the basic aspects and technology-oriented developments in this emerging discipline are covered by comprehensive and timely books. The series constitutes a survey of the relevant special topics, which are presented by leading experts in the field. These books will appeal to researchers, engineers, and advanced students.

Please view available titles in *NanoScience and Technology* on series homepage
http://www.springer.com/series/3705/

Stergios Logothetidis

Editor

Nanostructured Materials and Their Applications

With 134 Figures

Springer

Editor
Stergios Logothetidis
Aristotle University of Thessaloniki
Laboratory for Thin Films-Nanosystems and Nanometrology
Physics Department
Thessaloniki, Greece
logot@auth.gr

NanoScience and Technology ISSN 1434-4904
ISBN 978-3-642-22226-9 e-ISBN 978-3-642-22227-6
DOI 10.1007/978-3-642-22227-6
Springer Heidelberg Dordrecht London New York

Library of Congress Control Number: 2011942603

Printed on acid-free paper

Springer is part of Springer Science+Business Media (www.spinger.com)

Preface

Nanotechnology is one of the continuously emerging scientific areas combining knowledge from the fields of Physics, Chemistry, Biology, Medicine, Informatics and Engineering. Nanostructured materials and nanosystems are fabricated and fully characterised by nanotechnological tools and techniques, at sizes below 100 nm. Although there are restrictions related to nanoscale size, it is the handling and processing of matter at this scale that leads to the development of new and novel materials which may have the same bulk composition but widely varying properties. The diverse applications of nanomaterials ranging from electronic and engineering systems and devices, to optical and magnetic components, nanodevices in medicine, cosmetic merchandise, agricultural and food products are believed to pave the way and have a significant economical and societal impact.

This book gives an overview of nanostructures and nanomaterials applied in the fields of energy and organic electronics (*Chap. 1*). It combines the knowledge of advanced deposition and processing methods of nanomaterials, and state-of-the-art characterization techniques with special emphasis on the optical, electrical, morphological, surface and mechanical properties (mainly in *Chaps. 5 and 6*). Furthermore, it contains theoretical and experimental aspects for different types of nanomaterials, such as nanoparticles, nanotubes and thin films for organic electronics applications. Specifically it includes topics on carbon nanomaterials and nanotubes focusing on their different synthesis routes *(as shown in Chaps. 2 and 3)*, and full characterisation of their properties at a theoretical and experimental level for optoelectronics applications *(as shown in Chaps. 7–9)*. The different deposition techniques used to fabricate nanostructured thin films and the processing methods such as self-assembly and nanopatterning of surfaces are extensively described in *Chaps. 4 and 10.*

Thessaloniki
July 2011

Stergios Logothetidis

Contents

Contributors

M. Damnjanovic NanoLab, Faculty of Physics, POB 368, Belgrade 11001, Serbia, yqoq@rcub.bg.ac.rs

E. Glynos Institute for Materials and Processes, School of Engineering, University of Edinburgh, King's Buildings, Edinburgh EH9 3JL, UK, eglynos@umich.edu

Ch. Gravalidis Lab for Thin Films – Nanosystems and Nanometrology (LTFN), Physics Department, Aristotle University of Thessaloniki, 54124 Thessaloniki, Greece, cgrava@physics.auth.gr

Michael Heuken AIXTRON AG, Kaiserstr. 98, 52134 Herzogenrath, Germany, m.heuken@aixtron.com

S. Kassavetis Lab for Thin Films – Nanosystems and Nanometrology (LTFN), Physics Department, Aristotle University of Thessaloniki, 54124 Thessaloniki, Greece, skasa@physics.auth.gr

D. Keiper AIXTRON AG, Kaiserstr. 98, 52134 Herzogenrath, Germany, D.Keiper@aixtron.com

V. Koutsos Institute for Materials and Processes, School of Engineering, University of Edinburgh, King's Buildings, Edinburgh EH9 3JL, UK, vasileios.koutsos@ed.ac.uk

A. Laskarakis Lab for Thin Films – Nanosystems and Nanometrology (LTFN), Physics Department, Aristotle University of Thessaloniki, 54124 Thessaloniki, Greece, alask@physics.auth.gr

S. Logothetidis Physics Department, Lab for Thin Films – Nanosystems and Nanometrology (LTFN), Aristotle University of Thessaloniki, 54124 Thessaloniki, Greece, logot@auth.gr

Nico Meyer AIXTRON AG, Kaiserstr. 98, 52134 Herzogenrath, Germany, N.Meyer@aixtron.com

P. Patsalas University of Ioannina, Department of Materials Science and Engineering, 45110 Ioannina, Greece, ppats@cc.uoi.gr

K. Porfyrakis Department of Materials, University of Oxford, Parks Road, Oxford OX1 3PH, UK, kyriakos.porfyrakis@materials.ox.ac.uk

L. Tsetseris Department of Physics, National Technical University of Athens, 15780 Athens, Greece
Department of Physics, Aristotle University of Thessaloniki, 54124 Thessaloniki, Greece
Department of Physics and Astronomy, Vanderbilt University, Nashville, TN, USA, leont@mail.ntua.gr

A.C. Varonides Physics and EE Department, University of Scranton, Scranton, PA, USA, varonides@scranton.edu

J. Walker Institute for Materials and Processes, School of Engineering, University of Edinburgh, King's Buildings, Edinburgh EH9 3JL, UK, John.Walker@ed.ac.uk

J.H. Warner Department of Materials, University of Oxford, Parks Road, Oxford OX1 3PH, UK, jamie.warner@materials.ox.ac.uk

Chapter 1
Nanotechnology: Principles and Applications

S. Logothetidis

Abstract Nanotechnology is one of the leading scientific fields today since it combines knowledge from the fields of Physics, Chemistry, Biology, Medicine, Informatics, and Engineering. It is an emerging technological field with great potential to lead in great breakthroughs that can be applied in real life. Novel nano- and biomaterials, and nanodevices are fabricated and controlled by nanotechnology tools and techniques, which investigate and tune the properties, responses, and functions of living and non-living matter, at sizes below 100 nm. The application and use of nanomaterials in electronic and mechanical devices, in optical and magnetic components, quantum computing, tissue engineering, and other biotechnologies, with smallest features, widths well below 100 nm, are the economically most important parts of the nanotechnology nowadays and presumably in the near future. The number of nanoproducts is rapidly growing since more and more nanoengineered materials are reaching the global market The continuous revolution in nanotechnology will result in the fabrication of nanomaterials with properties and functionalities which are going to have positive changes in the lives of our citizens, be it in health, environment, electronics or any other field. In the energy generation challenge where the conventional fuel resources cannot remain the dominant energy source, taking into account the increasing consumption demand and the CO_2 emissions alternative renewable energy sources based on new technologies have to be promoted. Innovative solar cell technologies that utilize nanostructured materials and composite systems such as organic photovoltaics offer great technological potential due to their attractive properties such as the potential of large-scale and low-cost roll-to-roll manufacturing processes The advances in nanomaterials necessitate parallel progress of the nanometrology tools and techniques to characterize and manipulate nanostructures. Revolutionary new approaches in nanometrology

S. Logothetidis (✉)
Physics Department, Lab for Thin Films – Nanosystems & Nanometrology, Aristotle University of Thessaloniki, 54124 Thessaloniki, Greece
e-mail: logot@auth.gr

S. Logothetidis (ed.), *Nanostructured Materials and Their Applications*,
NanoScience and Technology, DOI 10.1007/978-3-642-22227-6_1,

will be required in the near future and the existing ones will have to be improved in terms of better resolution and sensitivity for elements and molecular species. Finally, the development of specific guidance for the safety evaluation of nanotechnology products is strongly recommended.

1.1 Introduction

The term nanotechnology comes from the combination of two words: the Greek numerical prefix nano referring to a billionth and the word technology. As an outcome, Nanotechnology or Nanoscaled Technology is generally considered to be at a size below 0.1 μm or 100 nm (a nanometer is one billionth of a meter, 10^{-9} m). Nanoscale science (or nanoscience) studies the phenomena, properties, and responses of materials at atomic, molecular, and macromolecular scales, and in general at sizes between 1 and 100 nm. In this scale, and especially below 5 nm, the properties of matter differ significantly (i.e., quantum-scale effects play an important role) from that at a larger particulate scale. Nanotechnology is then the design, the manipulation, the building, the production and application, by controlling the shape and size, the properties-responses and functionality of structures, and devices and systems of the order or less than 100 nm [1,2].

Nanotechnology is considered an emerging technology due to the possibility to advance well-established products and to create new products with totally new characteristics and functions with enormous potential in a wide range of applications. In addition to various industrial uses, great innovations are foreseen in information and communication technology, in biology and biotechnology, in medicine and medical technology, in metrology, etc. Significant applications of nanosciences and nanoengineering lie in the fields of pharmaceutics, cosmetics, processed food, chemical engineering, high-performance materials, electronics, precision mechanics, optics, energy production, and environmental sciences.

Nanotechnology is an emerging and dynamic field where over 50,000 nanotechnology articles have been published annually worldwide in recent years, and more than 2,500 patents are filed at major patent offices such as the European Patent Office [3].

Nanotechnology can help in solving serious humanity problems such as energy adequacy, climate change or fatal diseases: "Nanotechnology" Alcatel-Lucent is an area which has highly promising prospects for turning fundamental research into successful innovations. Not only to boost the competitiveness of our industry but also to create new products that will make positive changes in the lives of our citizens, be it in medicine, environment, electronics or any other field. Nanosciences and nanotechnologies open up new avenues of research and lead to new, useful, and sometimes unexpected applications. Novel materials and new-engineered surfaces allow making products that perform better. New medical treatments are emerging for fatal diseases, such as brain tumours and Alzheimer's disease. Computers are built with nanoscale components and improving their performance depends upon shrinking these dimensions yet further" [4].

Nanomaterials with unique properties such as: nanoparticles carbon nanotubes, fullerenes, quantum dots, quantum wires, nanofibers, and nanocomposites allow completely new applications to be found. Products containing engineered nanomaterials are already in the market. The range of commercial products available today is very broad, including metals, ceramics, polymers, smart textiles, cosmetics, sunscreens, electronics, paints and varnishes. However new methodologies and instrumentation have to be developed in order to increase our knowledge and information on their properties. Nanomaterials must be examined for potential effects on health as a matter of precaution, and their possible environmental impacts. The development of specific guidance documents at a global level for the safety evaluation of nanotechnology products is strongly recommended. Ethical and moral concerns also need to be addressed in parallel with the new developments.

Huge aspirations are coupled to nanotechnological developments in modern medicine. The potential medical applications are predominantly in diagnostics (disease diagnosis and imaging), monitoring, the availability of more durable and better prosthetics, and new drug-delivery systems for potentially harmful drugs. While products based on nanotechnology are actually reaching the market, sufficient knowledge on the associated toxicological risks is still lacking. Reducing the size of structures to nanolevel results in distinctly different properties. As well as the chemical composition, which largely dictates the intrinsic toxic properties, very small size appears to be a dominant indicator for drastic or toxic effects of particles. From a regulatory point of view, a risk management strategy is already a requirement for all medical technology applications [5–7].

In order to discuss the advances of nanotechnology in nanostructured materials, we presented first in Sect. 1.2 the methods and principles of nanoscale and nanotechnology, and the relevant processes. The impact of nanotechnology in the field of electronics is presented in Sect. 1.3. Energy harvesting and clean solar energy are presented in Sect. 1.4 focusing in a new emerging technology of plastic photovoltaics which is based on nanostructured materials. The techniques and the tools which are currently used to characterize and manipulate nanostructures are presented in Sect. 1.5. In Sect. 1.6, the future perspectives as well as the increasing instrumentational demands are discussed.

1.2 Methods and Principles of Nanotechnology

1.2.1 What Makes Nanostructures Unique

The use of nanostructured materials is not a recently discovered era. It dates back at the fourth century AD when Romans were using nanosized metals to decorate glasses and cups. One of the first known, and most famous example, is the Lycurgus

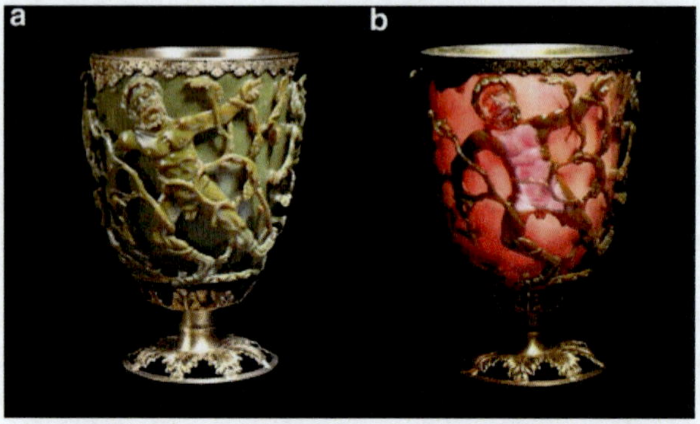

Fig. 1.1 The Lycurgus cup in reflected (**a**) and transmitted (**b**) light. Scene showing Lycurgus being enmeshed by Ambrosia, now transformed into a vine-shoot. Department of Prehistory and Europe, The British Museum. Height: 16.5 cm (with modern metal mounts), diameter: 13.2 cm. The Trustees of the British Museum [8]

cup (Fig. 1.1) [9], that was fabricated from nanoparticles (NPs) from gold and silver that were embedded in the glass. The cup depicts King Lycurgus of Thrace being dragged to the underworld. Under normal lighting, the cup appears green. However, when illuminated from within, it becomes vibrant red in color. In that cup, as well as in the famous stain glass windows from the tenth, eleventh, and twelfth centuries, metal NPs account for the visual appearance.

To shed light to the changes in visual appearance of gold, from the usual yellowish color to the reddish one that appears in the Lycurgus cup a comparison between differences of absorption spectra from a bulk gold metal film and a gold colloidal film (Fig. 1.2). The thin, bulk gold metal film absorbs across most of the visible part of the electromagnetic spectrum and very strongly in the IR and at all longer wavelengths. It dips slightly around 400–500 nm, and when held up to the light, such a thin film appears blue due to the weak transmission of light in this wavelength regime. On the contrary, the dilute gold colloid film displays total transparency at low photon energies (below 1.8 eV). Its absorption becomes intense in a sharp band around 2.3 eV (520 nm) This sharp absorption band is known as surface plasmon absorption band. Metals support SPs that are collective oscillations of excited free electrons and characterized by a resonant frequency. They can be either localized as for metal NPs or propagating as in the case of planar metal surfaces. Through manipulation of the geometry of the metallic structure, the SPR can be tuned depending on the application. The resonances of noble metals are mostly in the visible or near infrared region of the electromagnetic spectrum, which is of interest for decorative applications. Because of the plasmonic excitation of electrons in the metallic particles suspended within the glass matrix, the cup absorbs and scatters blue and green light – the relatively short wavelengths of the visible spectrum. When viewed in reflected light, the plasmonic scattering gives the cup a greenish hue, but if a white light source is placed within the goblet, the glass

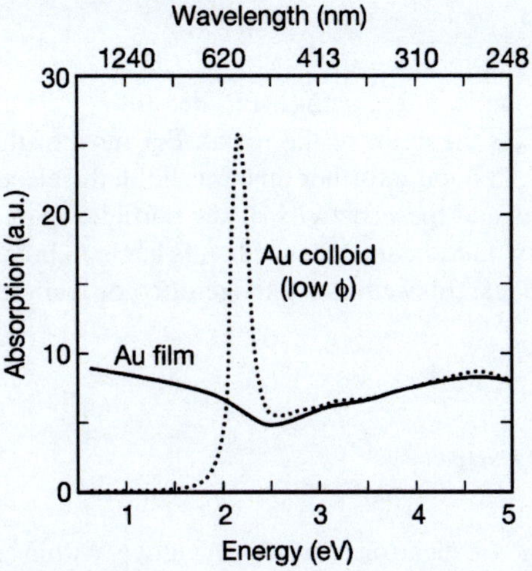

Fig. 1.2 Absorption spectra of a gold nanocrystal film which absorbs only above 1.8 eV like a semiconducting material due to the quantum confinement effect and a thin, bulk gold metal film of equivalent thickness which absorbs like a typical metal in the infrared energy region. ϕ is the volume fraction of gold in the sample [10]

appears red because it transmits only the longer wavelengths and absorbs the shorter ones [10].

1.2.2 Size Dependence

The aforementioned ability of gold as well as of other noble metals and semiconductors relies on quantum confinement which is a very successful model for describing the size dependent electronic structure of nanometer sized materials According to this theory electrons are confined in all three dimensions causing matter to behave completely different in terms of its optical and electronic properties. When the dimension of a material approaches the electron wavelength in one or more dimensions, quantum mechanical characteristics of the electrons that are not manifest in the bulk material can start to contribute to or even dominate the physical properties of the material [11].

Besides quantum size effects, the nanomaterials behavior is different due to surface effects which dominate as nanocrystal size decreases. Reducing the size of a crystal from 30 to 3 nm, the number of atoms on its surface increases from 5% to 50% beginning to perturb the periodicity of the "infinite" lattice. In that sense, atoms at the surface have fewer direct neighbors than atoms in the bulk and as a result they are less stabilized than bulk atoms [11]. The origin of the quantum size effects strongly depends on the type of bonding in the crystal.

1.2.3 Metal NPs

For metals, the electron mean free path (MFP) determines the thermal and electrical conductivity and affects the color of the metal. For most of the metals, MFP is of the order of 5–50 nm. Reducing further this threshold, the electrons begin to scatter off the crystal surface, and the resistivity of the particles increases. For very small metal particles, the conduction and valence bands begin to break down into discrete levels. For gold particles, this causes a change in color from red to orange at sizes around 1.5 nm.

1.2.4 Quantum Dots

In a bulk semiconductor electrons can freely move within an area from a few nanometers to a few hundred of nanometers as defined by the Bohr radius. Thus continuous conduction and valence energy bands exist which are separated by an energy gap. Contrary, in a quantum dot, where excitons cannot move freely, discrete atomic like states with energies that are determined by the quantum dot radius appear.

The effect of quantum confinement has a great technological interest from semiconductors and optoelectronics to biological applications. As depicted in Fig. 1.3, by changing the particle size the emitting color of quantum dots can be

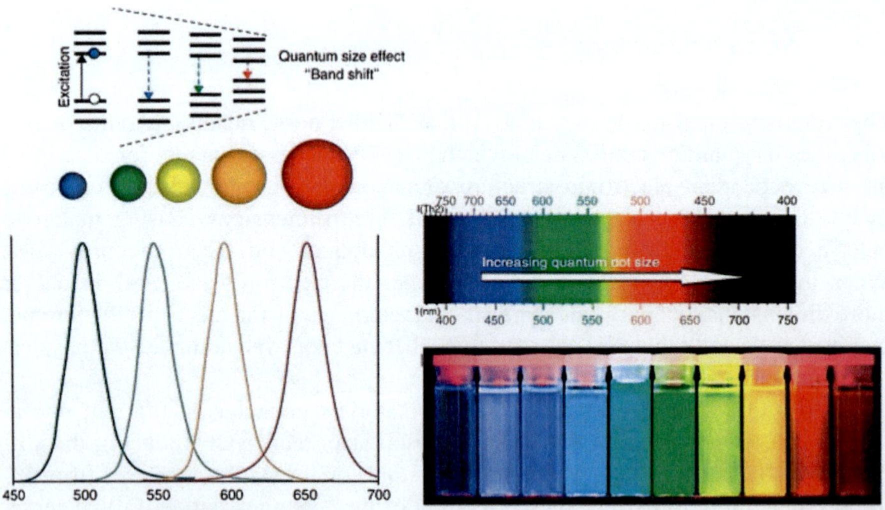

Fig. 1.3 Schematic drawing representing the changes on optical behavior of nanoparticles associated with their size. *Top*: Electronic structure of QDs with "blue shift" due to quantum confinement [12]

tuned. Shorter quantum dots emit shorter wavelength of light and bigger quantum dots emits longer wavelengths of light. The energy band gap E_g is correlated with size: as the dimension of particles decreases, the energy increases.

$$E_g^{Qd} = E_g^b + \left(\frac{h^2}{8R^2}\right)\left(\frac{1}{m_e} + \frac{1}{m_h}\right) - \left(\frac{1.8e^2}{4\pi\varepsilon_0\varepsilon R}\right),$$

where $E_{g,b}$ and $E_{g,QD}$ are the bandgap energies of the bulk solid and quantum dot, respectively, R is the quantum dot radius, m_e is the effective mass of the electron in the solid, "e" is elementary charge of the electron, "h" is Planck's constant, the m_h is the effective mass of the hole in the solid, and " ε" is the dielectric constant of the solid. The middle term on the right-hand side of the above equation is a 'particle-in-a-box-like' term for the exciton, while the third term on the righthand side of the equation represents the electron–hole pair Coulombic attraction, mediated by the solid [12]. Some of new applications of quantum dots are memories, transistors, detectors, and lasers and quantum computers.

1.2.5 Nanotechnology Imitates Nature

When a droplet of water lands on the lotus leaf, it beads up, rolls off the leaf surface without leaving a trace of water behind, and washes away any dirt along its way. This self-cleaning property fascinated scientists for a long time until recently, when scientists realized that this peculiar behavior is due to the nanostructures present on the surface of the lotus leaf. They term this as super-hydrophobicity. These can be integrated in numerous parts of the building infrastructure. New developing nanostructured surfaces behave like the lotus leaf and stay dry when water lands on them. Such degree of water repellency exceeds even that of one of the most well-known hydrophobic materials, polytetrafluoroethylene (PTFE) or Teflon [13].

The natural technology of gecko foot-hairs can provide biological inspiration for future design of remarkably effective adhesives. Since gravity plays a negligible role at nanoscale, the van der Waals forces become very important. The van der Waals energy per unit area, E/α, between two infinite parallel surfaces is:

$$\frac{E}{a} = -\frac{A}{12\pi D^2},$$

where A is the Hamaker constant which is a constant that depends on the material properties (and can be positive or negative depending on the intervening medium) and D is the distance between the surfaces. The feet of a Gekko gecko contains approximately one billion spatulae that can provide a sufficiently large surface area in close contact with the substrate for adhesion to be the result of van der Waals forces [14] (Fig. 1.4).

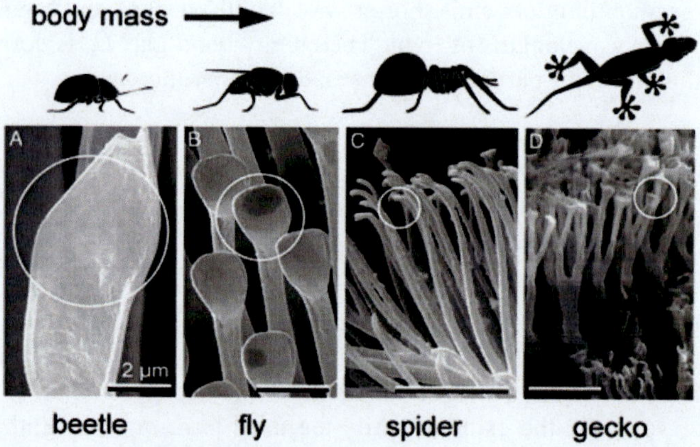

Fig. 1.4 Terminal elements (*circles*) in animals with hairy design of attachment pads. Note that heavier animals exhibit finer adhesion structures [15]

Different methods for the synthesis of nanoengineered materials and devices can accommodate precursors from solid, liquid, or gas phases and encompass a tremendously varied set of experimental techniques. A detailed presentation of these is beyond the scope of this review. In general, however, most synthetic methods can be classified into two main approaches: "*top-down*" and "*bottom-up*" approaches and combinations of them (Fig. 1.5). "*Top-down*" (photolithography, microcontact printing) techniques begin with a macroscopic material or group of materials and incorporate smaller-scale details into them, whereas "*bottom-up*" (organic-synthesis, self-assembly) approaches begin by designing and synthesizing custom-made molecules that have the ability to self-assemble or self-organize into higher order mesoscale and macroscale structures. Bottom-up approach aims to guide the assembly of atomic and molecular constituents into organized surface structures through processes inherent in the manipulated system [16].

One example of the bottom-up approach is self-assembly. Self-assembly is the fundamental principle which generates structural organization on all scales from molecules to galaxies. It is a method of integration in which the components spontaneously assemble, until a stable structure of minimum energy is reached. Furthermore, self-assembly is not limited to nanoscaled molecules but can be carried out on just any scale, making it a powerful bottom-up method for Nanotechnology. Self-assembly of colloidal nanoparticles on surfaces is extensively discussed in Chap. 10 by Koutsos et al. An alternative example of bottom-up approach uses scanning probe microscopes to position molecules at the desired position on surfaces.

One the most common self-assembled monolayers (SAMs) preparation methodology is that of alkanethiols on gold which was first reported in 1983 by Nuzzo and Allara [17]. The preparation of SAMs typically involves immersing a gold-coated substrate in a dilute solution of the alkanethiol in ethanol as shown Fig 1.6. A monolayer spontaneously assembles at the surface of the substrate over the next

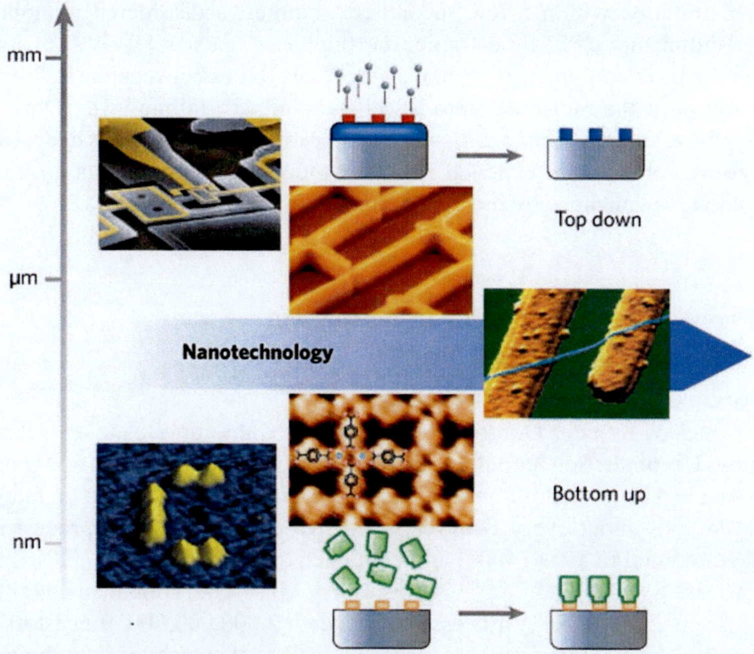

Fig. 1.5 Two approaches to control matter at the nanoscale: shown (clockwise from top) are an electron microscopy image of a nanomechanical electrometer obtained by electron-beam lithography, patterned films of carbon nanotubes obtained by microcontact printing and catalytic growth, a single carbon nanotube connecting two electrodes, a regular metal–organic nanoporous network integrating iron atoms and functional molecules, and seven carbon monoxide molecules forming the letter "C" positioned with the tip of a scanning tunneling microscope [16]

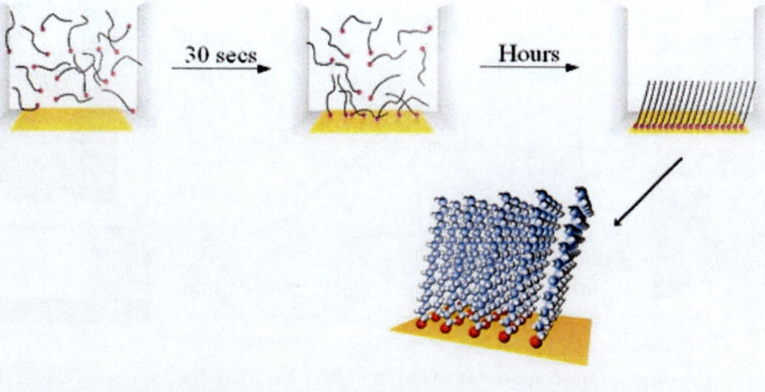

Fig. 1.6 Schematic representation of the self-assembly process. Initially alkanethiols come down onto the gold surface. As more alkanethiols come to the surface, the layer begins to organize and pack into an ordered monolayer [18]

1 to 24 h. Initially, within a few seconds to minutes, a disordered monolayer is formed. Within this early time frame, the thickness reaches 80–90% of its final value. As the layer continues to form, van der Waals forces between the hydrocarbon chains help pack the molecules into a well-ordered, crystalline layer. During this ordering phase, contaminants are displaced (for example, adventitious hydrocarbons on the gold), solvents are expelled from the monolayer, and defects are reduced while packing is enhanced by increased packing of the alkanethiols [18].

1.3 From Microelectronics to Nanoelectronics and Molecular Electronics

In 1965, Intel co-founder Gordon Moore forecasted the rapid pace of technology innovation. His prediction, popularly known as "Moore's Law," states that transistor density, that is the number of transistors in an integrated circuit or chip on integrated circuits, doubles about every two years (Fig. 1.7). The first microprocessor was introduced by Intel in 1971 (4004) and contained 2,300 transistors. In 2004, Intel's fastest processor (Intel® Itanium® 2 processor, 9 MB cache) contained 592,000,000 transistors. In 2010 Intel's processor exceeded 2,000,000,000 transistors [20]. However, this development is now reaching a wall so that smaller is no longer any faster. The prime reason for the limitation the semiconductor electronics experiences is its power dissipation and thus heat [21].

Figure 1.8 shows the evolution miniaturization of the conducting channel between the two other contacts, the source and the drain of a transistor. The channel length which is made of n- or p-doped silicon was reduced from 50 nm in 2003 to 10 nm today (2011). However technical factors limit the top-down development of microelectronics, the non-scalability of the MOS transistor below a critical size

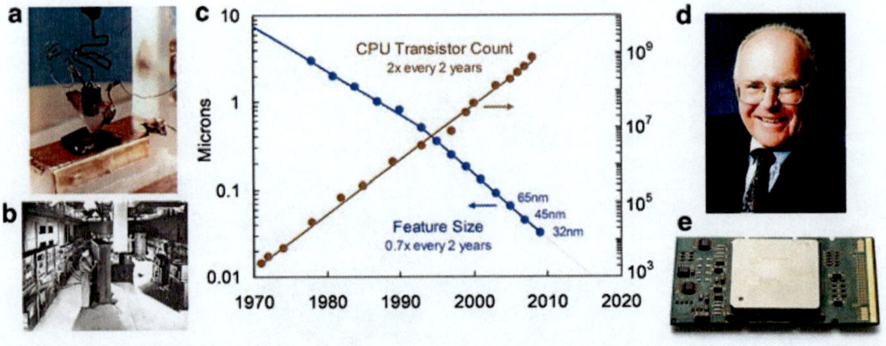

Fig. 1.7 (**a**) The first, point-contact transistor invented by John Bardeen and Walter Brattain in December 1947. Photo courtesy of Alcatel-Lucent. (**b**) Electronic Numerical Integrator and Computer, ENIAC conceived and designed by John Mauchly and J. Presper Eckert in 1946. (**c**) Transistors per microprocessor history [19]. (**d**) Gordon Moore. (**e**) Intel Itanium 2 processor

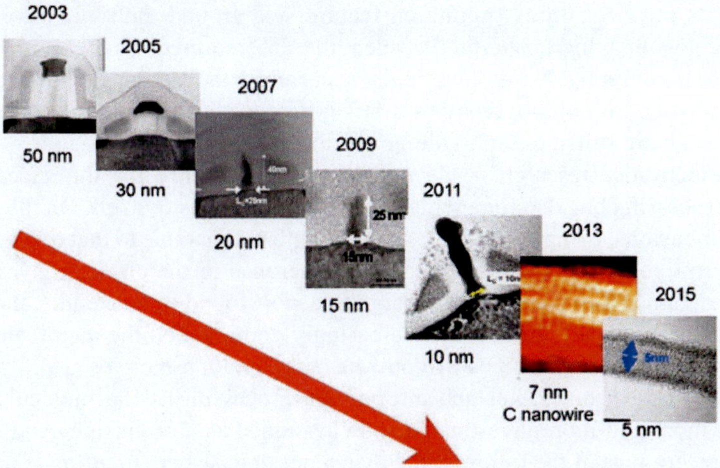

Fig. 1.8 Scaling history of the transistor channel length. The channel length was reduced from 50 nm in 2003 to 10 nm today. Future perspectives incorporate the use of nanowires in OFETs [20]

L_{phys} (the physical limit) or the impossibility of batch defining, using proximity masks, feature sizes below a critical one L_{litho} (the lithographic limit) [22]. At the IC (integrated circuit) level, the channel length (L) of the transistor is very critical because as this length decreases we have: (a) increase in the number of the transistors of the IC, thus then more "logic gates", so more "processing" power and (b) decrease in the "responding time" of the "logical operations". At the transistor level, the channel length (L) of the transistor is included in the basic equation of the gain factor β.

$$\beta = \frac{\mu_\varepsilon}{t_{\text{OX}}}\left(\frac{W}{L}\right).$$

That means, when the length L decreases, we have: (a) increase in the gain factor β of the transistor and (b) better directivity of electrons in the channel path [23].

Organic field-effect transistors (OFETs) are an alternative technology with high technological potential due to the possibility of low-cost and large-area manufacturing processes. An OFET uses an organic semiconductor in its channel and can be prepared either by vacuum evaporation of small molecules or by solution-casting of polymers or small molecules. It has been demonstrated that single-walled carbon nanotubes (SWNTs) can be used as quasi-one-dimensional (1D) electrodes to construct organic FETs with molecular-scale width (~2 nm) and channel length (down to 1–3 nm) [24].

Ultra-dense integrated circuits with features smaller than 10 nm would provide enormous benefits for all information technologies, including computing, networking, and signal processing. The top-down route of the silicon technology has indeed been relatively easy to run until its basic step (optical lithography) has met its physical limits (minimum feature size around the light wavelength). To overcome this limit, shorter wavelengths are required, such as in extreme ultraviolet lithography. X-ray lithography is currently the leading technology in the drive to replace photolithography as a large-scale production tool because it uses masks, which are suited to high-volume production.

Nanoelectronics research is currently looking not only for the successor to CMOS processing but also for a replacement for the transistor itself. On the scale of 10 nm dimensions, components have a wavelength comparable to that of an electron at the Fermi energy. The confinement and coherence of the electron gives rise to gross deviations from the classical charge transport found in conventional devices. Quantum-mechanical laws become increasingly dominant on the nanoscale, and it is probable that nanoelectronics will operate on quantum principles [25].

Molecular electronics, i.e., the information processing at the molecular-scale, becomes more and more investigated and envisioned as a promising candidate for the nanoelectronics of the future. More than a possible answer to ultimate miniaturization problem in nanoelectronics, molecular electronics is foreseen as a possible way to assemble a large number of nanoscale objects (molecules, nanoparticles, nanotubes, and nanowires) to form new devices and circuit architectures [26].

The difference between molecular- (nano) and micro-electronics is not the size (dimensionality), but the profoundly different device- and system-level solutions, the device physics, and the phenomena, fabrication, and topologies/organizations/architectures. Three-dimensional topology molecular and nanoelectronic devices, engineered from atomic aggregates and synthesized utilizing *bottom-up* fabrication, exhibit quantum phenomena and electrochemomechanical effects that should be uniquely utilized. Given technological advancements, molecular electronics proponents believe purposeful bottom-up design will be more efficient than the top-down method, and that the incredible structural diversity available to the chemist will lead to more effective molecules, thus approaching optional functionality for each application. A single mole of molecular switches, weighing about 450 g and synthesized in small reactors (a 22-L flask might suffice for most steps of the synthesis), contains 6×10^{23} molecules – a number greater than all the transistors ever made. While we do not expect to build a circuit in which each single molecule is both addressable and connected to a power supply (at least not in the first few generations), the extremely large numbers of switches available in a small mass illustrate one reason molecular electronics can be a powerful tool for future computing development [27].

1.4 Nano in Energy and Clean Energy

Energy is one of the most challenging needs of humanity, and is highest on the list of priorities and requisites for human welfare [28]. According to the International Energy Agency (IEA), World's primary energy demand will increase by 36% between 2008 and 2035. Electricity demand is expected to grow by 2.2% per year between 2008 and 2035. Taking in account the CO_2 emissions and the global climate change impact on life and the health of the planet renewable energy sources will have to play a central role in moving the world onto a more secure, reliable, and sustainable energy path [29].

Solar energy is the most abundant, inexhaustible and clean of all the renewable energy resources till date. The power from sun intercepted by the earth is about 1.8×10^{11} MW, which is many times larger than the present rate of all the energy consumption. Photovoltaic technology is one of the finest ways to harness the solar power [30].

Figure 1.9 shows the history of confirmed "champion" laboratory cell efficiencies. The performance of conventional solar cells is approaching a plateau; only incremental improvements have been accomplished in the last decade despite dedicated R&D effort.

Tandem solar cells based on III–V materials have achieved the highest efficiencies of any present photovoltaic device exceeding 40% recently However, the cost of these devices is very high, limiting their application to space applications [32]. The efficiencies reached with commercial solar cell modules are significantly lower than those of the best laboratory cells due to losses incurred during scaleup. The typical size of "champion laboratory cells" is in the square centimeter range or even below, facilitating the collection of photocurrent. High efficiency multijunction solar cells with finely-tuned quantum wells are presented in Chap. 5 by Varonides.

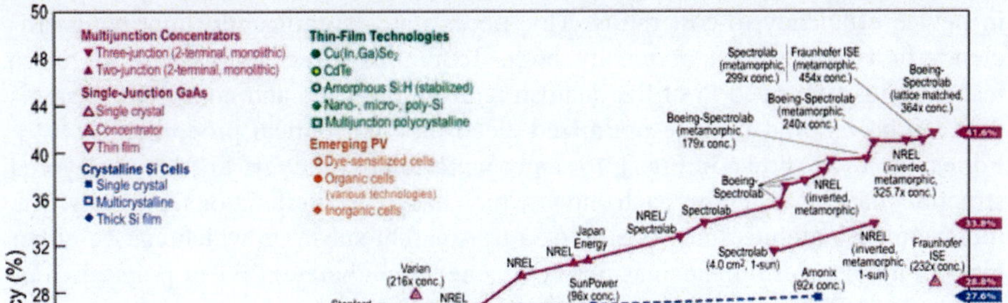

Fig. 1.9 Historic summary of champion cell efficiencies for various PV technologies. Tandem solar cells based on III–V materials present the highest efficiencies of any present photovoltaic device exceeding 40% [31]

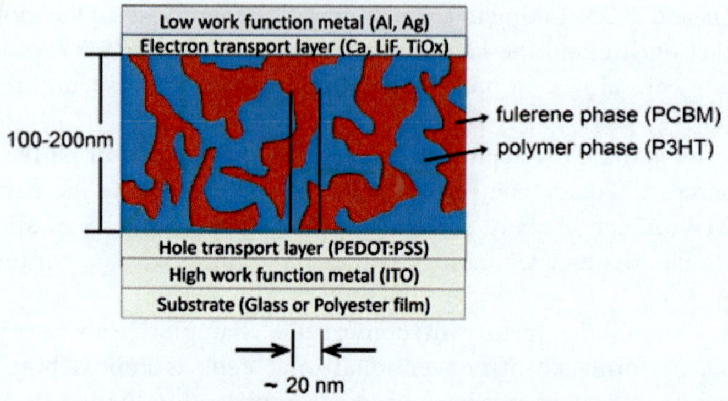

Fig. 1.10 Schematic structure of a typical organic solar cell device

Although inorganic semiconductors (silicon, amorphous silicon, gallium arsenide, and sulfide salts) have been the primary focus, the photosensitivity and the photovoltaic effects in devices made with organic materials have also been explored because of the advantages such as the potential of large-scale and low-cost roll-to-roll manufacturing processes. More specific, organic photovoltaics offer great technological potential as a renewable energy source due to their mechanical flexibility, low weight of plastic materials and easy thin-film casting technology [33–37].

Plastic electronics technologies are aimed at producing significant improvements in device efficiency-to-cost ratios. This necessitates significantly improving efficiency or reducing cost or ideally both. To realize these goals, many of these technologies will need to utilize nanostructured materials and composite systems that can be tailored to have optimized electronic and optical properties [38, 39]. For example, as shown in Fig. 1.10, an organic solar cell consist of a multilayered structure made of thin films each one of which has a certain functional property. The most common architecture consists of a transparent substrate which can be either glass or a polyester film such as poly(ethylene terephthalate) PET or poly(ethylene naphthalate) PEN. A highwork function metal electrode such as indium tin oxide (ITO) serves as an anode for collecting holes and a lowwork function metal such as aluminum serves as the cathode collecting the electrons which are produced in the active layer. Additional buffer layers such as hole transport layers (PEDOT:PSS) or electron transport layers (Ca, LiF, and TiO_x) are placed between the electrodes and the active layer to provide better energy level alignment and better ohmic contact between the organic layer and the metal electrodes.

The most successful active layer up to date consists of a bulk heterojunction (BHJ) that is formed by a p-type semiconductor (electron donor), such as poly(3-hexylthiophene) (P3HT) with an n-type semiconductor (electron acceptor), such as methanofullerene derivatives (PCBM). Due to the low dielectric constant in organic components ($\sim$3) photoexcitation leads to a strongly bound exciton, which needs to be dissociated into free carriers. This dissociation can take place in a strong electric field or at the donor–acceptor interface. Then, free carriers need to be transported to the corresponding electrodes via drift and diffusion processes, where they are collected, giving rise to an electric current. The morphology of the active donor–acceptor film is critical for charge generation and transport, strongly influencing device performance. Despite the rapid increase in device performance observed recently, much effort is still required to understand the fundamental processes of photovoltaic energy generation, in particular to elucidate the complex relationship between nanoscale morphology/electronic properties and device performance and to further develop the appropriate nanometrology needed to address this interplay [40].

1.5 Nanotechnology Tools: Nanometrology

The great development in Nanotechnology has given birth to the need of knowing of the dimensions that characterize its nanostructure. This lead to the appearance of a new scientific field called Nanometrology. Nanometrology is the science and practice of measurement of functionally important, mostly dimensional parameters and components with at least one critical dimension which is smaller than 100 nm. Success in nanomanufacturing of devices will rely on new nanometrologies needed to measure basic materials properties including their sensitivities to environmental conditions and their variations, to control the nanofabrication processes and materials functionalities, and to explore failure mechanisms. In order to study and explore the complex nanosystems, highly sophisticated experimental, theoretical, and modeling tools are required [41]. Especially, the visualization, characterization, and manipulation of materials and devices require sophisticated imaging and quantitative techniques with spatial and temporal resolutions on the order of 10^{-6} and below to the molecular level. In addition, these techniques are critical for understanding the relationship and interface between nanoscopic and mesoscopic/macroscopic scales, a particularly important objective for biological and medical applications [42]. The need for better characterization at the nanoscale derives from the correlation between the macroscopic functional properties with the nanoscale structural characteristics of nanomaterials which is a prerequisite for the development of emerging low-cost manufacturing technological fields such as organic electronics. These include organic solar cells (OPVs), organic light emitting diodes (OLEDs) and organic field-effect transistors (OFETs), and others. Insights on the nanomorphology as well as the conduction mechanisms at the various interfaces that exist in these multilayered devices are crucial for the development of the plastic electronic technology and the construction of better products. Examples of important tools available at the moment include highly focused synchrotron X-ray sources and related techniques that provide detailed molecular structural information by directly probing the atomic arrangement of atoms; scanning probe microscopy that allow three-dimensional-type topographical atomic and molecular views or optical responses of nanoscale structures; in situ optical monitoring techniques that allow the monitoring and evaluation of building block assembly and growth; optical methods, with the capability of measuring in air, vacuum, and in liquid environment for the study of protein and cells adsorption on solid surfaces, they have been employed to discriminate and identify bacteria at the species level and it is very promising for analytical purposes in biochemistry and in medicine [43–45].

The nanometrology methods need measurements that should be performed in real-time to allow simultaneous measurement of properties and imaging of material features at the nanoscale. These nanometrology techniques should be supported by physical models that allow the de-convolution of probe–sample interactions as well as to interpret sub-surface and interface behaviors. *Ellipsometry* is a key-technique meeting the aforementioned demands. It can be applied during the nanofabrication processes and provide valuable information concerning the optical, vibrational, structural, and morphological properties, the composition as well as the thickness and the mechanisms of the specimen under growth or synthesis conditions in nanoscale. Further correlation between optical and other physical properties can lead to a more complementary characterization and evaluation of materials and devices [46–49]. Additional information about the possibilities and application of this technique are given in Chap. 7 by Laskarakis et al. *X-ray photoelectron spectroscopy (XPS)* is one of the most quantitative techniques to determine both atomic concentration and the chemical environment of the species at the surface of a sample. XPS has a high potential for non-destructive depth profiling (<10 nm from the surface) in angle resolved mode or by using synchrotron radiation for variable excitation energy XPS. *Secondary ion mass spectroscopy (SIMS)* is a highly sensitive technique that may be used to determine the composition of a material, typically at or near the surface. Deduced from specific calibrations, a quantity of atoms as a function of their mass charge ratios is measured as a function of depth. Detection limits for trace elements are typically between 10^{12} and 10^{16} atoms per cm^3 [50]. *X-ray reflectivity (XRR)* is also a powerful tool for investigating monolithic and multilayered film structures. It is one of the few methods that, with great accuracy, not only allows information on the free surface and the interface to be extracted but also the mass density and the thickness of very thin film of the order of a few nanometers along the direction normal to the sample surface to be determined. XRR is able to offer accurate thickness determination for both homogeneous thin films and multilayers with the same precision, as well as densities, surface, and interface roughness of constituent layers. In addition, other promising nanometrology techniques include the *tip-enhanced-Raman-spectroscopy (TERS)* [51–53]. TERS combines the capabilities of Raman spectroscopy that has been used for many years for the single layer and even single molecule detection in terms of its chemical properties, with the advantages of an atomic force microscopy tip that is put close to the sample area that is illuminated by the Raman laser beam. In this way, a significant increase in the Raman signal and in the lateral resolution by up to nine orders of magnitude takes place. Thus, TERS can be used for the chemical analysis of very small areas and for the imaging of nanostructures as well as of other materials such as proteins and biomolecules (Fig. 1.11) [56, 57].

Another important nanometrology method is *nanoindentation* that has rapidly become the method of choice for quantitative determination of the mechanical properties (as hardness and elastic modulus) of thin films and small volumes of material. In Fig. 1.12a, b the nanoindentation load–displacement curve coming from nanoindentation test to amorphous carbon thin film grown on silicon (001) substrate and the imprint of the Berkovich diamond indenter on the surface of aluminum is presented. More information about this technique are provided in Chap. 6 by Kassavetis et al.

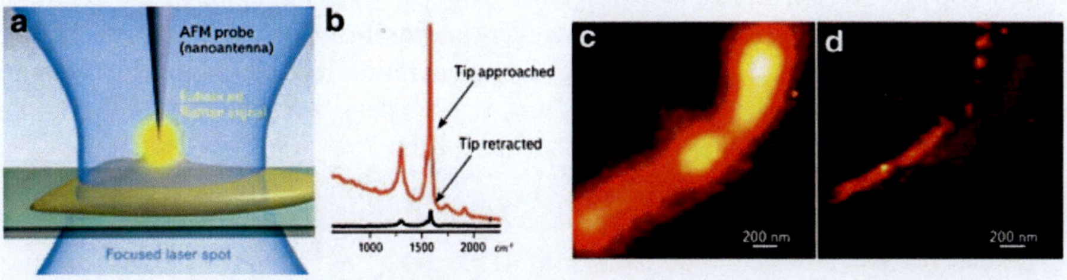

Fig. 1.11 (**a**) A specially prepared AFM probe (metalcoated cantilever or etched metal wire is precisely positioned inside a tightly focused laser spot. (**b**) Intensity of carbon nanotube G- and D-Raman bands increases by several orders of magnitude when the special AFM probe is landed and positioned over a small (5 nm height) nanotube bundle – the effect of Tip Enhanced Raman Scattering (TERS). (**c**) "Conventional" confocal Raman image of the nanotube bundle, the observed width of the bundle is ∼250 nm (diffraction limit of confocal microscopy, laser wavelength – 633 nm). (**d**) TERS image of the same bundle – now the observed width is ∼50 nm. In this example, TERS provides more than four-times better spatial resolution as compared to confocal microscopy [54, 55]

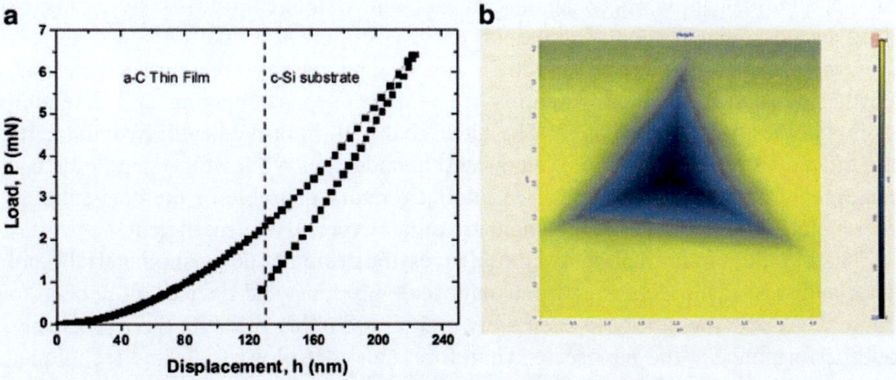

Fig. 1.12 (**a**) Nanoindentation load–displacement curve and (**b**) the imprint of the Berkovich type diamond indenter on aluminum

Finally, *scanning probe microscopes (SPMs)* are standard instruments at scientific and industrial laboratories that allow imaging, modifications, and manipulations with the nanoobjects. They permit imaging of a surface topography and correlation with different physical properties within a very broad range of magnifications, from millimeter to nanometer-scale range. Atomic force microscopy (AFM) and AFM related techniques (e.g., scanning near-field optical microscopy – SNOM) have become sophisticated tools, not only to image surfaces of molecules but also to measure molecular forces between molecules. This is substantially increasing our knowledge of molecular interactions. Electrical scanning probe microscopy (SPM) techniques have already made a number of important contributions to the field of organic electronics. Conductive atomic force microscopy (c-AFM), electrostatic force microscopy (EFM), scanning Kelvin probe microscopy (SKPM), and similar variants were successfully used to elucidate charge injection/extraction, transport, trapping, and generation/recombination in organic devices [58].

1.6 Future Perspectives

Nanotechnology is distinguished by its interdisciplinary nature. As investigations at the nanolevel are occurring in a variety of fields, it is expected that the results of this research are going to have a significant impact on a broad range of applications [59] Nanomaterials with tailored unique properties have limitless possibilities in materials science Products where the addition of a relatively small amount of functionalized nanoparticles or carbon nanotubes leads to a major change in the properties are going to revolutionize many commercial technologies.

It is believed that nanotechnology can greatly contribute to the evolution of modern medical approaches and practices. Nanoscale constructs are already used in therapeutical applications against cancer and pathogens mostly by acting as drug carriers. Also either in-vivo or ex-vivo engineered scaffolds and tissues are implanted in patients whose own organs and tissues are damaged or lost. Furthermore, specific nanostructures are widely used as imaging and detection agents in diagnostic procedures. The ideal goal is to improve health by enhancing the efficacy and safety of nanosystems and nanodevices while at the same time use nanomedicine in order to cure diseases that remain incurable or the conventional therapeutical approaches against them are either expensive or inefficient.

The advances in fundamental nanosciences, the design of new nanomaterials, and ultimately the manufacturing of new nanoscale products and devices all depend to some degree on the ability to accurately and reproducibly measure their properties and performance at the nanoscale. Therefore, nanometrology tools and techniques are both integral to the emerging nanotechnology enterprise and are two of the main areas critical to the success of nanotechnology. Decades of nanoscience research have led to remarkable progress in nanotechnology as well as an evolution of instrumentation and metrology suitable for some nanoscale measurements.

Consequently, today's suite of metrology tools has been designed to meet the needs of exploratory nanoscale research. New techniques, tools, instruments and infrastructure will be needed to support a successful nanomanufacturing industry.

The currently available metrology tools are also beginning to reach the limits of resolution and accuracy and are not expected to meet future requirements for nanotechnology or nanomanufacturing. Novel methods and combinations, such as the TERS technique, achieve much higher resolution values since they provide a significant increase in the Raman signal and in the lateral resolution by up to nine orders of magnitude. This combination overcomes the difficulties that originate from low signal since the Raman systems have limit in lateral resolution of 300 μm and require high laser power for surface investigation because the measured Raman intensity is six orders of magnitude lower than the excitation power. Thus, TERS is a promising technique and we can see it in the near future to be used for probing the chemical analysis of very small areas and for the imaging of nanostructures and biomolecules such as proteins. New approaches have to be developed and existing ones based on XPS, X-ray absorption spectroscopy, SPM and SIMS have to be improved in terms of better spectral and spatial resolution, better contrast and better sensitivity for elements and molecular species. Ideally new methods should have capabilities to work in situ, at ambient air and/or in liquid surroundings.

However, clever new approaches need to be developed. For this, it is required to understand the fundamental mechanisms by which the probes of the nanometrology measuring systems interact with the materials and objects that are being measured. Also, it is important to develop standard samples and to construct standardized procedures for measurements in nanometer scale, which enable the transfer of the properties and response of the unit from the nanometer to macroscopic scale without any appreciable loss of accuracy, for certifying, calibrating, and checking nanometrology instruments. Finally, even with the vast array of current tools available, the important question is whether or not they are providing the required information or reams of inconsequential data. Revolutionary approaches to the nanometrology needed may be required in the near future and therefore, revolutionary and not just evolutionary instrumentation and metrology are needed.

1.7 Summary

Nanotechnology is an emerging technology with applications in several scientific and research fields, such as information and communication technology, electronics, energy, biology, medical technology, etc. Novel nano- and biomaterials, and nanodevices are fabricated and controlled by nanotechnology tools and techniques, which investigate and tune the properties, responses and functions of living and non-living matter, at sizes below 100 nm.

Nanotechnology is a science with huge potential and great expectations. The daily announcements of new discoveries and breakthroughs are going to influence all aspects of human society. Nanomaterials bring new possibilities by tailoring the optical, the electronic the mechanical, the chemical, and the magnetic properties. In the last few years there was a rapid progress in the fabrication and processing of nanostructures. As a result nanophase materials and applications are already in the market and a large volume of new applications is expected over the next several years. However, the development and commercialization of products containing nanomaterials raises many of the same issues as with introduction of any new technology, including concerns about the toxicity and environmental impact of nanomaterial exposures. Despite the extensive research of the last decade the literature on toxicological risks of the application of nanotechnology in medical technology is scarce.

In order to investigate in depth the complex nanosystems, highly sophisticated nanoscale precision metrology tools are required. The advances in nanomaterials necessitate parallel progress of the nanometrology tools and techniques. Examples of important nanometrology tools as they have been discussed above include: ellipsometry, highly focused X-ray sources and related techniques, nanoindentation and scanning probe microscopies. The above described nanometrology methods contribute towards the understanding of several aspects of the state-of-the-art nanomaterials in terms of their optical, structural, and nanomechanical properties. The nanoscale precision and the detailed investigation that these nanometrology techniques offer, give them an enormous potential for even more advanced applications for the improvement of the quality of research and of the everyday life.

References

1. K.D. Sattler, *Handbook of Nanophysics, Principles and Methods* (CRC, New York, 2010)
2. B. Bhushan, *Handbook of Nanotechnology* (Springer, Berlin, 2004)
3. C. Huang, A. Notten, N. Rasters, J. Technol. Transf. **36**, 145–172 (2011)
4. F. Simonis S. Schilthuizen, *Nanotechnology Innovation Opportunities For Tomorrow's Defence* (TNO Science & Industry, 2006)
5. K. Park, Nanotechnology: what it can do for drug delivery, perspective. J. Control. Release **120**, 1–3 (2007)
6. W.H. de Jong, B. Roszek, R.E. Geertsma, Nanotechnology in medical applications: possible risks for human health. RIVM report 265001002, 2005 (RIVM, National Institute for Public Health and the Environment, Bilthoven, 2005)
7. Nanotechnology, biotechnology, information technology & cognitive science – NBIC developments
8. I. Freestone, N. Meeks, M. Sax, C. Higgitt, Gold Bull. **40**(4),270 (2007)
9. British Museum, Lycurgus cup, http://www.britishmuseum.org
10. P. Mulvaney et al., MRS Bull. **26**, 1009 (2001)
11. A.G. Davies, J.M.T. Thompson, Advances in Nanoengineering Electronics, Materials and Assembly. Royal Society Series on Advances in Science **3**, Imperial College Press, London, 2007, Page 3 (Introduction, G. Davies, School of Electronic and Electrical Engineering, University of Leeds, Leeds, UK).
12. H.S. Mansur, Wiley Interdiscip. Rev. Nanomed. Nanobiotechnol. **2**, 113–129 (2010)
13. http://www.hyphotech.com
14. K. Autumn, Y.A. Liang, S.T. Hsieh, W. Zesch, W.P. Chan, T.W. Kenny, R. Fearing, R.J. Full, Nature **405**, 681–685 (2000)

15. E. Arzt, S. Gorb, R. Spolenak, From micro to nano contacts in biological attachment devices. Proc. Natl. Acad. Sci. U.S.A. **100**(19), 10603–10606 (2003)
16. J.V. Barth, G. Costantini, K. Kern, Nature **437**, 671–679 (2005)
17. R.G. Nuzzo, D.L. Allara, J. Am. Chem. Soc. **105**, 4481–4483 (1983)
18. Asemblon, Inc., Self-Assembling Molecules, http://www.asemblon.com/
19. Moore's Law past 32 nm: Future Challenges in Device Scaling, Kelin Kuhn/IWCE/Beijing/ 2009
20. http://www.intel.com
21. S. Anders, M.G. Blamire, F.I.m. Buchholz, D.G. Crétéd, R. Cristiano, P. Febvre, L. Fritzsch, A. Herr, E. Il'ichev, J. Kohlmann, J. Kunert, H.G. Meyer, J. Niemeyer, T. Ortlepp, H. Rogalla, T. Schurig, M. Siegel, R. Stolz, E. Tarte, H.J.M. ter Brake, H. Toepfer, J.C. Villegier, A.M. Zagoskin, A.B. Zorin, Physica C **470**, 2079–2126 (2010)
22. G.F. Cerofolini et al., Microelectron. Eng. **81**, 405–419 (2005)
23. N. Weste, K. Eshraghian, *Principles of CMOS VLSI Design: A Systems Perspectives* (Addison-Wesley, Boston, 1993)
24. P. Qi, A. Javey, M. Rolandi, Q. Wang, E. Yenilmez, H. Dai, J. Am. Chem. Soc. **126**, 11774–11775 (2004)
25. J.K. Gimzewski, Nanoelectronics, in McGraw-Hill Yearbook of Science and Technology 2000 (McGraw-Hill, New York, 1999), 274–278
26. D. Vuillaume, Molecular nanoelectronics, Proc. IEEE **98**, 12, 2111–2123 (2010)
27. S.E. Lyshevski *Nano and Molecular Electronics Handbook* (CRC, New York, 2007)
28. A.F. Ghoniem, Prog. Energy Combust. Sci. **37**, 15–51 (2011)
29. The International Energy Agency, World Energy Outlook, http://www.worldenergyoutlook.org/
30. B. Parida, S. Iniyan R. Goic, Renew. Sustain. Energy Rev. **15**, 1625–1636 (2011)
31. National Renewable Energy Laboratory, NREL, http://www.nrel.gov/solar/
32. M. Grätzel, Phil. Trans. R. Soc. A **365**, 993–1005 (2007)
33. G. Yu, J. Gao, J.C. Hummelen, F. Wudl, A.J. Heeger, Polymer photovoltaic cells: enhanced efficiencies via a network of internal donor–acceptor heterojunctions. Science **270**, 1789–1795 (1995)
34. S. Logothetidis, Mater. Sci. Eng. B **152**, 96–104 (2008)
35. R.J. Hamers, Flexible electronic futures. Nature **412**, 489 (2001)
36. F. Padinger, R.S. Rittberger N.S. Sariciftci, Adv. Funct. Mater. **13**(2) (2003)
37. S.R. Forrest, The path to ubiquitous and low cost organic electronic appliances on plastic. Nature **428**, 911 (2004)
38. S.E. Shaheen, D.S. Ginley, Photovoltaics for the Next Generation: Organic-Based Solar Cells, Dekker Encyclopedia of Nanoscience and Nanotechnology, Schwarz, Contescu, and Putyera, Eds.; Marcel Dekker, Inc.: New York, 2879–2895 (2004)
39. C. Contescu K. Putyera, *Dekker Encyclopedia of Nanoscience and Nanotechnology* 2nd edn. (CRC, New York, 2009)
40. P.G. Nicholson, F.A. Castro, Nanotechnology **21,** 492001 (2010)
41. K. Nomura, H. Ohta, A. Takagi, T. Kamiya, M. Hirano H. Hosono, Nature **432**, 488 (2004)
42. D.J. Whitehouse, *The Handbook of Surface and Nanometrology* (Taylor & Francis, New York, 2002)
43. V. Karagkiozaki, S. Logothetidis, N. Kalfagiannis et al, Atomic force microscopy probing platelet activation behavior on titanium nitride nanocoatings for biomedical applications. Nanomed. Nanotechnol. Biol. Med. **5**(1), 64–72 (2009)
44. S. Lousinian, N. Kalfagiannis, S. Logothetidis, Albumin and fibrinogen adsorption on boron nitride and carbon-based thin films. Mater. Sci. Eng. B 152 **1–3**, 12–15 (2008)
45. S. Lousinian, S. Logothetidis, In-situ and real-time protein adsorption study by spectroscopic ellipsometry. Thin Solid Films **516**, 8002–8008 (2008)
46. R.M.A. Azzam N.M. Bashara, *Ellipsometry and Polarized Light* (North Holland, New York, 1977)

47. S. Logothetidis, *Thin Films Handbook: Processing, Characterization and Properties* (Academic, New York, 2001) p. 227
48. L.V. Keldysh, D.A. Kirzhnits, A.A. Maradudin, *The Dielectric Function Of Condensed Systems* (North-Holland, New York)
49. Y. Kitano, Y. Kinoshita, T. Ashida, Morphology and crystal-structure of an a axis oriented, highly crystalline poly(ethylene-terephthalate). Polymer **36**, 10, 1989 (1995)
50. R.K. Leach, R. Boyd, T. Burke, H.U. Danzebrink, K. Dirscherl, T. Dziomba, M. Gee, L. Koenders, V. Morazzani, A. Pidduck, D. Roy, W.E.S. Unger and A. Yacoot, Nanotechnology **22**, 062001 (15.(2011)
51. K.J. Yi, X.N. He, Y.S. Zhou et al., Tip-enhanced near-field Raman spectroscopy with a scanning tunneling microscope and side-illumination optics. Rev. Sci. Instrum. **79**(7) 073706 (2008)
52. J. Steidtner, B. Pettinger Tip-enhanced Raman spectroscopy and microscopy on single dye molecules with 15 nm resolution. Phys. Rev. Lett. **100**, 236101 (2008)
53. M. Motohashi, N. Hayazawa, A. Tarun et al, Depolarization effect in reflection-mode tip-enhanced Raman scattering for Raman active crystals. J. Appl. Phys. **103**(3), 034309 (2008)
54. http://www.ntmdt.com/
55. S.S. Kharintsev, G.G. Hoffmann, P.S. Dorozhkin, G. de With J. Loos, Nanotechnology **18**, 315502 (2007)
56. W. Zhang, B.S. Yeo, T. Schmid et al., Single molecule tip-enhanced Raman spectroscopy with silver tips. J. Phys. Chem. C **111**(4) 1733–1738 (2007)
57. B.S. Yeo, T. Schmid, W. Zhang et al., Towards rapid nanoscale chemical analysis using tip-enhanced Raman spectroscopy with Ag-coated dielectric tips. Anal. Bioanal. Chem. **387**(8) 2655–2662 (2007)
58. L.S.C. Pingree, O.G. Reid D.S. Ginger, Adv. Mater. **21**, 19–28 (2009)
59. J. Kennedy in *The Yearbook of Nanotechnology and Society*, ed. by E Fisher, C Selin JM. Wetmore Nanotechnology: the Future is Coming Sooner Than You Think, Yearbook of Nanotechnology in Society vol 1 (Springer, New York, 2008), pp. 1–21

Chapter 2
Carbon Nanomaterials: Synthesis, Properties and Applications

Kyriakos Porfyrakis and Jamie H. Warner

2.1 Introduction

It is widely acknowledged that nanosciences and nanotechnologies are going to play an important role in our society and that they have the potential to create benefits in many technological areas including materials science, information technology as well as energy and the environment. The word "nano" is of Greek origin and means dwarf. The name reveals that we are dealing with materials and processing at fundamental length scales. Conventionally, nanotechnology involves processes and materials at length scales between 100 nm and 1 nm or below (1 nanometer is equal to 10^{-9} m).

Carbon nanomaterials are remarkable materials at the lower end of this scale. The fullerene C_{60} (or Buckyball) is 0.7 nm across, while single-walled carbon nanotubes (SWNTs) have typically diameters between 1 and 2 nm. To put this into perspective, the Earth, a football and a C_{60} molecule all have approximately spherical shape. However the Earth is about one hundred million times larger than a football, which in turn is about one hundred million times larger than C_{60}. It is really astonishing that we are able to manipulate and visualise molecules with nearly atomic precision. In this chapter we shall take a closer look at the properties and technological potential of carbon nanomaterials such as fullerenes, carbon nanotubes and graphene.

K. Porfyrakis (✉) · J.H. Warner
Department of Materials, University of Oxford, Parks Road, Oxford, OX1 3PH, UK
e-mail: kyriakos.porfyrakis@materials.ox.ac.uk; jamie.warner@materials.ox.ac.uk

S. Logothetidis (ed.), *Nanostructured Materials and Their Applications*, NanoScience and Technology, DOI 10.1007/978-3-642-22227-6_2,

2.2 Fullerenes and Their Derivatives

The story of fullerenes as chemical structures started in the imagination of the insightful chemist E. Osawa [1] in 1970. Fifteen years later Kroto, Heath, OBrien, Curl and Smalley published their seminal paper on the discovery of C_{60} [2]. That work caused worldwide sensation and led to the award of Nobel Prize in chemistry. Initially fullerenes were produced in tiny quantities, which prevented their in-depth study. The synthetic breakthrough came in 1990. Krätschmer, Lamb, Fostiropoulos and Huffman were the first to produce macroscopic quantities of C_{60} by resistive heating of graphite rods under a He atmosphere [3]. This breakthrough led to an explosion of scientific research. Fullerenes became available in sufficient amounts for spectroscopic analysis and further chemistry. The most characteristic spectroscopic analysis comes from the ^{13}C-NMR spectrum. Figure 2.1 shows the spectrum of C_{60} in a $CS_2/CDCl_3$ mixture. Due to the curvature of the molecule, C_{60} has got the character and behaviour of a polyalkene with 60 equivalent carbon atoms of sp^2 hybridisation. Hence only one peak appears in the NMR spectrum at $\approx$143 ppm chemical shift.

C_{70} on the other hand has an elongated "rugby-ball" shape with D_{5h} symmetry. The ^{13}C-NMR spectrum of C_{70} is shown in Fig. 2.2.

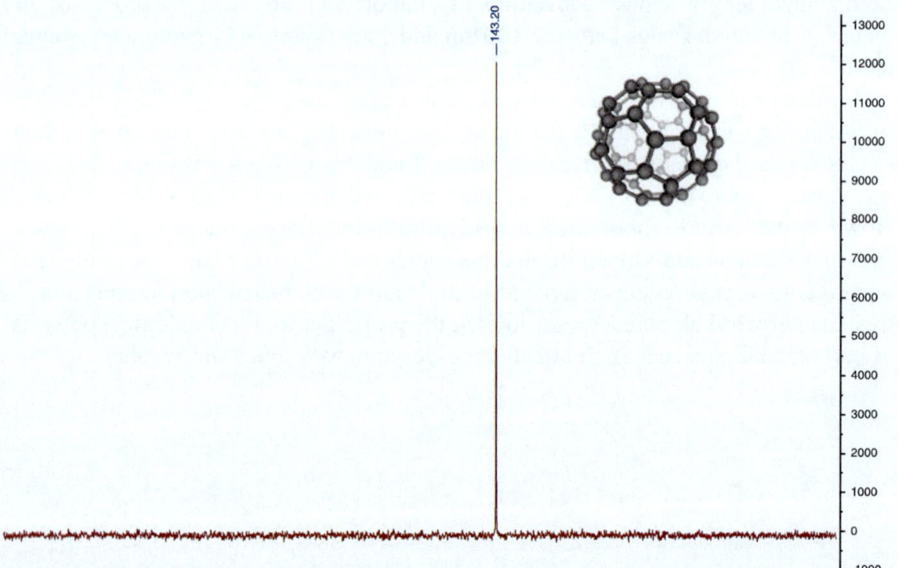

Fig. 2.1 ^{13}C-NMR spectrum of C_{60} in a $CS_2/CDCl_3$ mixture. The molecule has got icosahedral symmetry I_h. The inset shows a model of the C_{60} molecule

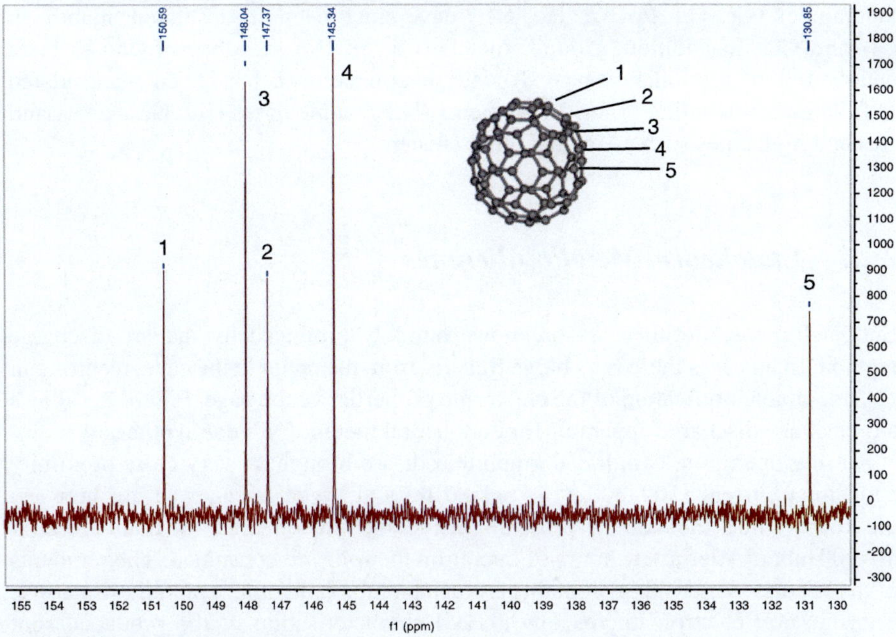

Fig. 2.2 ^{13}C-NMR spectrum of C_{70} in a $CS_2/CDCl_3$ mixture. The molecule has got D_{5h} symmetry. The inset shows a model of the C_{70} molecule

It can be seen that there are five peaks at 150.6, 148, 147.4, 145.3 and 131 ppm chemical shift. These shifts correspond to the five types of carbon atoms present at the molecule (from pole to equator), as shown in the inset. The intensity ratio of these peaks 1:2:1:2:1 corresponds to the relevant abundancy of the carbon atoms. In addition to NMR, a lot of information on the electronic structure and physicochemical properties of fullerenes has been obtained by other spectroscopies such as UV-Vis, FTIR and Raman. The more interested reader can find out more about these techniques and their application to fullerene science in the relevant bibliography and references therein [4].

2.2.1 *Synthesis of Endohedral Fullerenes*

Undoubtedly the most important feature of fullerene molecules is their cage-like structure. These are molecules with an enclosed interior space. It was not for long that chemists started trapping atoms inside the fullerene empty "shell". Fullerenes containing atoms or clusters in their interior are called *endohedral* fullerenes. The first endohedral metallofullerenes were lanthanum containing fullerene cages, produced by vaporisation of lanthanum-doped graphite rods. The most stable lanthanofullerene was found to be $La@C_{82}$. Other group-3 metals (Sc, Y) and lanthanides (Ce, Gd, Pr, Nd, Ho, etc.) have since been encapsulated, mainly in C_{82} and C_{80}. In addition, group-2 metals (Ca, Sr, Ba) have been found to form endohedral metallofullerenes [5]. To date several elements have been encapsulated in fullerenes, including group-15 elements (N,P), noble gases (He, Ne, Ar, Kr and Xe) and molecules such as H_2 and other clusters.

2.2.2 *Endohedral Metallofullerenes*

Endohedral metallofullerenes are now routinely produced by the arc-discharge method. In all cases there is a charge transfer from the metal to the cage, resulting in considerable modification of the electronic properties of the cage. Figure 2.3 shows a typical arc-discharge apparatus for endohedral metallofullerene synthesis.

During operation, two doped graphite rods are brought in very close proximity and direct current (100–300 A) is passed through them forming an arc between the rods, while the helium pressure inside the arc chamber is maintained at 40–100 mbar. After a few hours of operation the rods are consumed. The remnants of the vaporised rods (slug) contain carbon nanotubes and other graphitic structures. Transmission electron microscopy (TEM) characterisation of the produced soot shows that it too is comprised mainly from amorphous carbon and graphitic structures. More importantly, the soot contains typically 10–20% fullerenes. The yield of fullerenes via the arc-discharge method is very sensitive to parameters such as He pressure, current and rod size. One discerning feature of the arc discharge shown in Fig. 2.3 is the ability to collect *and* dissolve the produced soot in an organic solvent, such as toluene and carbon disulphide, in anaerobic conditions to avoid unnecessary degradation of the endohedral metallofullerenes. The fullerenes are separated from the insolubles by soxhlet extraction. The fullerene extract is filtered, re-dissolved in fresh solvent and then passed through a high performance liquid chromatography (HPLC) apparatus in order to separate the individual fullerene species.

A particular class of endohedral metallofullerenes that has witnessed a blossoming interest over the past few years are the so-called TNTs (trimetallic nitride templated) or cluster fullerenes. They were first reported by Stevenson et al. in 1999 [6]. They were synthesised by introducing a small amount of nitrogen into the arc-discharge reactor. Although they were incorrectly identified as "small bandgap" materials initially, they were found to be very stable and quite abundant in the reactor. They can be produced with yields of 5% or higher and with a small number of isomers. Since then a large family of these materials has been synthesised with the structure $M_3N = C_{2n}$ ($34 \leq n \leq 44$) where M is a metal atom or a combination of up to three different metal atoms [7]. Some of them were even found to break the isolated pentagon rule (IPR), which normally dictates the stability of fullerene molecules [8]. $Er_3N = C_{80}$ exhibits photoluminescence from Er^{3+} ions at 1.5 μm, a wavelength region attractive for telecommunications [9]. This property gives it technological value. Comprehensive reviews on these materials can be found elsewhere [10, 11].

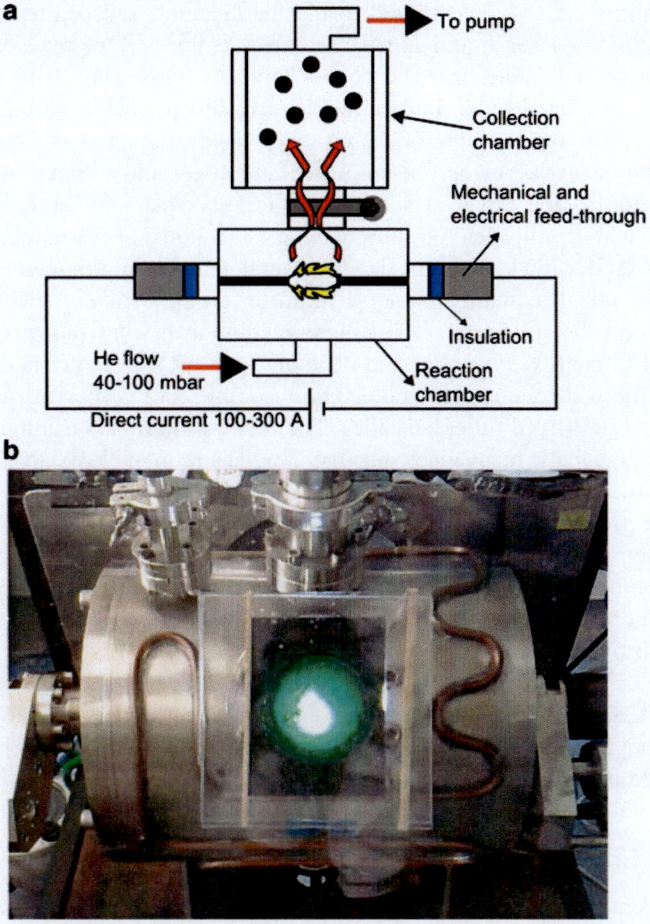

Fig. 2.3 (**a**) Schematic illustration of an arc-reactor for the production of endohedral metallofullerenes. Two doped graphite rods are brought in close proximity and high current is passed through them. An electric arc forms and the rods begin to evaporate. The soot that is produced is carried by helium to the collection chamber where the soot condenses on the liquid nitrogen-cooled walls. (**b**) Picture of the arc-discharge apparatus during operation

2.2.3 *Endohedral Nitrogen Fullerenes*

In addition to metallofullerenes, non-metals such as nitrogen and phosphorus have also been encapsulated in fullerenes. In contrast to metallofullerenes, these atoms appear to be more stable in smaller cages such as C_{60} and C_{70}. N@C_{60} and N@C_{70} are produced using the ion implantation method developed by Weidinger and co-workers at the Hahn-Meitner Institut in Germany [12]. Approximately 1 or 2 g of C_{60} are put into an effusion cell inside a vacuum chamber evacuated at a pressure of 10^{-6} mbar or lower. The effusion cell is heated at around 500°C. Under these conditions the C_{60} is sublimed inside the chamber and begins to condense onto a water-cooled (or liquid nitrogen-cooled) copper target placed above the effusion cell. At the same time the copper target is bombarded with low energy nitrogen ions produced by an ion source. Best results are achieved using a mass-separating source, for example one producing N+ preferentially to N_2+. Typical values for the beam energy and beam current are 40 eV and 1–3 mA, respectively. The orientation of the target is such that it is located at 45° angle to both the effusion cell and the nitrogen ion source. After a few hours of operation, the copper target is covered with a fullerene layer, several tens of micrometers thick. The copper target is subsequently immersed into an organic solvent such as CS_2 in order to extract the fullerenes. The fullerene solution is ultrasonicated for a few minutes and filtered. Between 60 and 70% of N@C_{60}/C_{60} mixture is dissolved in CS_2, while the rest remains insoluble. The insoluble soot comprises polymerised fullerenes and destroyed fullerene cages. The filtered solution is examined by EPR (electron paramagnetic resonance) spectroscopy. The ratio of N@C_{60}/C_{60} is 10^{-4} to 10^{-5}.

The same group developed an alternative method of producing N@C_{60}: the glow discharge method [13]. This is a rather simpler experimental set-up compared to the ion implantation device. A quartz tube is equipped with two water-cooled copper electrodes at opposite ends. The chamber is filled with low pressure (approximately 0.1 mbar) nitrogen gas. High voltage (of the order of 1 kV) is applied across the electrodes resulting in ionisation of the nitrogen gas. At the same time, several tens of grams of C_{60} are put inside the quartz tube. The whole apparatus is then inserted in a tube oven and the system is heated up to 500°C. C_{60} sublimes and is exposed to the ionised nitrogen gas before condensing on the copper electrodes. At the end of the operation the copper electrodes are immersed in organic solvents and the produced N@C_{60}/C_{60} mixture is extracted. The yield of the glow discharge method is 10^{-5} to 10^{-6} in terms of the N@C_{60}/C_{60} ratio.

More recently N@C_{60} has been produced using an electron cyclotron resonance (ECR) plasma source with a yield approaching 3×10^{-4} under optimised conditions [14]. However it was not clear whether this method can be used for scaled-up production of N@C_{60} or even if the amounts produced are comparable with the two more established methods described earlier.

2.2.4 *Molecular Synthesis of Endohedral Fullerenes*

The most impressive synthetic work on endohedral fullerene synthesis stems from the group of Komatsu and Murata at Kyoto University in Japan. They developed the "molecular surgery" approach, in which the fullerene cage is opened, then doped with an atom or molecule and finally closed using a series of organic chemistry reactions [15, 16]. They created a 13-membered circular ring orifice on the cage with a sulphur atom on its rim. Then they managed to insert molecular hydrogen through the orifice into the cage with a 100% yield. Finally they used a series of reductive coupling and annealing steps in order to close the orifice and hence produce $H_2 = C_{60}$. This is a challenging but an elegant method that may be extended in other types of endohedral fullerenes as well. Certainly $D_2 = C_{60}$ and $HD_2 = C_{60}$ seem possible as do other small atoms or molecules that can be inserted in fullerene cages in this way.

2.2.5 *Purification of Endohedral Fullerenes*

In nearly all cases, the synthesis of endohedral fullerenes is only the first step towards acquiring high purity individual species. Multi-stage HPLC is the established method for fullerene isolation. This is usually the most crucial and laborious step in the whole process. A combination of state-of-the-art chromatography columns tailored for fullerene purification is required for the complete isolation of isomerically pure fullerenes. Fullerenes tend to elute with size, thus C_{60} is the first one to elute followed by C_{70} and the larger cage fullerenes, including endohedral fullerenes. When fullerenes are synthesised by the arc discharge process, C_{60} accounts for about 60% of the total fullerene production whereas C_{70} represents approximately 25% of the production. The remainder 15% comprises larger empty cages as well as endohedral fullerenes. Three or four stages of HPLC through a suite of reverse-phase columns is usually enough to isolate a few milligram of high purity endohedral species [17–19]. This process may sound complicated. However it is routine compared to the purification of N@C_{60} and related species. The two main obstacles are, first, the very low yields of the N@C_{60} production methods and, second, the fact that C_{60} and N@C_{60} are chemically almost identical. Nevertheless two groups managed independently to completely isolate N@C_{60} and N@C_{70} with a purity of higher than 99.5% via a combination of mutliple injections and recycling HPLC through an appropriate column (such as the Cosmosil 5-PBB by Nacalai Tesque) [20–22].

2.2.6 *Properties and Applications*

It comes as no surprise that the presence of incarcerated atom(s) changes the physico-chemical and electronic properties of the cage. These molecules have been found to have unusual properties such as paramagnetism [23], luminescence [24] and non-linear optical response [25]. Endohedral metallofullerenes containing gadolinium ions have been proposed for use as MRI contrast agents [26]. Their magnetic relaxivity values are higher than the relaxivities of clinically used Gd(III) chelates; hence Gd metallofullerene compounds can serve as good relaxation agents for water protons. In vivo studies of holmium-containing metallofullerenes have demonstrated their potential as radiotracers for biomedical applications [27]. Recently it has been shown that water-soluble derivatives of endohedral fullerenes can inhibit the growth of malignant tumors in vivo [28]. The mechanism of such effect is not quite clear but the authors speculate that this behaviour relates with the capacity of endohedral fullerenes to "scavenge" reactive oxygen species.

Apart from biomedical applications, another area where fullerenes in general have shown tremendous potential is optoelectronics. Empty cage fullerenes have already been used successfully in organic photovoltaic devices with an efficiency reaching 5% in polymer bulk-heterojunction devices [29]. Endohedral metallofullerenes have higher electron affinity ($\approx$3 eV) and lower ionisation potential ($\approx$7 eV) than empty cages; hence they are better electron acceptors and better electron donors. The first published work on using endohedral fullerene derivatives as electron acceptors in solar cell devices appeared very recently [30]. Ross et al. produced devices with power conversion efficiency $\geq$4%. This is quite encouraging and it looks likely that endohedral fullerenes will play a prominent role in photovoltaic cells in the near future. An additional advantage is that one can tune the HOMO-LUMO gap and the electronic properties of endohedral fullerenes by selective encapsulation of specific metal ion(s) combinations. For example $Sc = C_{82}$ has ionisation potential of 6.45 eV while $La = C_{82}$ has ionisation potential of 6.19 eV.

2.2.6.1 Endohedral Fullerenes for Quantum Information Processing

One characteristic property of endohedral fullerenes is the presence of electron spin on the molecule. The electron spin (and indeed the nuclear spin too) is a quantum property; hence quantum information can be embodied in the electron/nuclear spin state. Sc-, Y- and La-metallofullerenes have unpaired electrons [5]. The unpaired electron spin resides mostly on the cage [31]. Nitrogen containing fullerenes also carry quantum information embodied in the electron spin of the unpaired electrons of the nitrogen atom. In this case the spin-density is almost entirely localised inside the carbon cage [32]. The atomic nitrogen orbitals fit "snuggly" within the fullerene and because of the curvature of the cage, interaction with the carbon orbitals is not favourable [33]. The relative isolation of the electron spin from the environment makes these systems attractive for quantum computation schemes. For successful realisation of quantum computing there must be adequate immunity to decoherence: the degrading of quantum states due to interactions with the environment. Provided the coherence time is sufficiently long compared with the gate operation time, fault-tolerant error correction schemes can be implemented to overcome decoherence [34]. To date, NMR spin systems have hosted the most complex quantum algorithms [35]. In these systems the quantum bits, or qubits, are embodied in the slowly decohering nuclear spins of the atoms of a molecule. However, owing to the fact that the thermal energy is always large compared to the nuclear Zeeman energy in NMR experiments, NMR-based quantum computers face a fundamental limitation in scalability and appear to be practically limited to around ten qubits. Since scalability is one of the preconditions of effective quantum computation [36, 37], the practical applications of NMR-based quantum computers seem limited. EPR offers the potential to use experimentally accessible fields and temperatures to approximate pure quantum states. Endohedral fullerenes are molecular materials; therefore they are all identical at the most fundamental level.

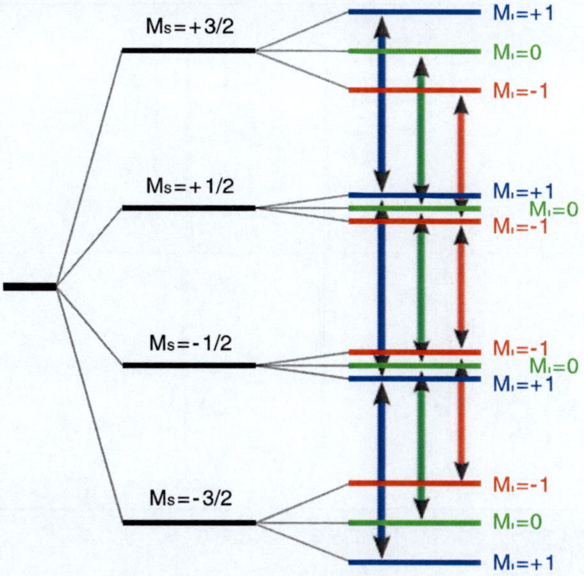

Fig. 2.4 Energy level diagram of $^{14}N@C_{60}$ in a magnetic field. $^{14}N@C_{60}$ has electron spin $S = 3/2$ and nuclear spin $I = 1$. This gives rise to a 12-level structure due to the Zeeman splitting. Taking into account just the first-order hyperfine interaction, the allowed transitions (the selection rules are $\Delta M_S = 1$ and $\Delta M_I = 0$) are triply degenerate

In addition, sophisticated chemistry can be applied to create scalable nanostructures based on these molecules.

2.2.6.2 N@C$_{60}$ as a Spin Qubit

$N@C_{60}$ is a $S = 3/2$ electron spin system coupled to the ^{14}N nuclear spin $I = 1$ via an isotropic hyperfine interaction. This gives rise to the rich energy level diagram shown in Fig. 2.4.

Taking into account only the first-order hyperfine interaction, the nine allowed electron transitions are triply degenerate. For this reason the observed continuous-wave EPR spectrum of $N@C_{60}$ dissolved in CS_2 at room temperature (shown in Fig. 2.5) comprises three sharp resonance peaks.

The three EPR resonances are quite narrow. Their intrinsic linewidth was measured to be $\leq 0.3\ \mu T$. In fact the linewidth is mainly limited by the resolution of the spectrometer and in particular the magnet stability and field homogeneity. The ability of $N@C_{60}$ to store quantum information effectively is demonstrated by the spin-lattice relaxation time T_1 and the phase-coherence time T_2. T_2 was measured to be ≥ 0.25 ms in CS_2 solution at 160 K [38]. Pulse sequences in a typical EPR spectrometer are of the order of 30 ns. This corresponds to more than 10^4 electron spin Rabi oscillations, before decoherence occurs. These properties of the

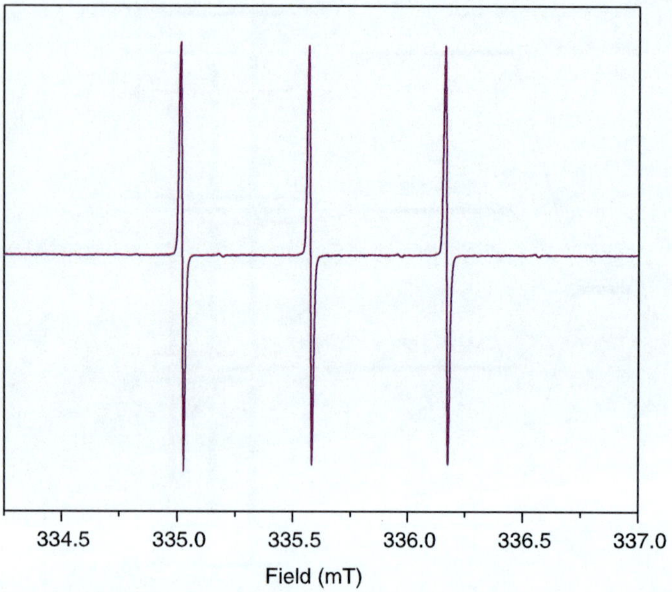

Fig. 2.5 Continuous-wave EPR spectrum of $^{14}N@C_{60}$ in a CS_2 solution. The two small resonances on either side of the central peak are associated with ^{15}N nuclei naturally abundant (less than 0.4%) in the sample

$N@C_{60}$ system ensure that it meets all the basic criteria for fault-tolerant quantum computing. Consequently $N@C_{60}$ has been proposed as a building bock of a solid-state quantum computer [39–41].

2.2.7 *Chemistry of Endohedral Fullerenes*

Most of the practical applications highlighted above require the construction of controlled molecular arrangements. For instance, in the case of quantum information processing, the smallest device where universal quantum gates could be applied is a two-qubit system. This requirement translates into linking two endohedral molecules together via covalent or non-covalent bonds. Moreover, dipolar coupling between adjacent spins is proportional to $1/r^3$, where r is their spatial separation. Hence, in order to control the strength of the spin-spin coupling, one must control their spatial separation. In other words, chemistry can be used to control the coupling strength of the qubits.

The chemistry of fullerenes is already well established. For example, Diels-Alder cycloaddition, Bingel, Prato and other reaction schemes have been employed in the synthesis of fullerene adducts and a rich relevant literature exists [42, 43]. The situation is markedly different when it comes to endohedral fullerenes. There are two main obstacles on the road to chemical functionalisation of endohedral

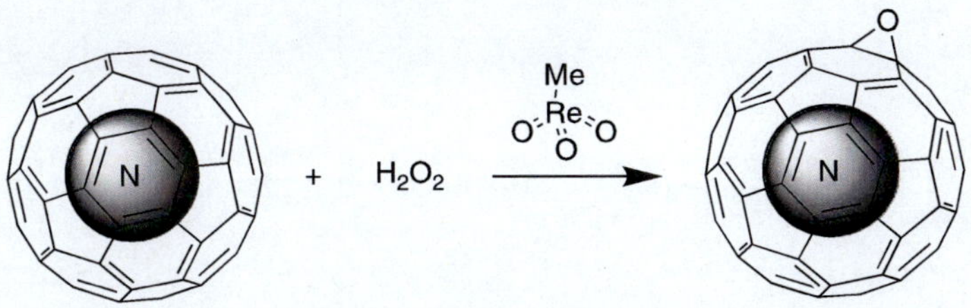

1

2

Scheme 1: The epoxidation of **1** (N@C_{60}) to form **2** (N@C_{60}O)

fullerenes. The first one is the difficulty in producing these materials in multi-milligram quantities. The second (and an equally formidable one) is the lower thermal and photolytic stability of some functionalised endohedrals such as N@C_{60} adducts. The first derivative of N@C_{60} was N = $C_{61}(COOEt)_2$, produced via the Bingel reaction of a N@C_{60}/C_{60} mixture with diethyl bromomalonate [44]. At the time no effect on the stability of N = $C_{61}(COOEt)_2$ was reported.

It was not until a few years later that another functionalisation of N@C_{60} was published. Franco et al. reported a series of fulleropyrrolidine derivatives of N@C_{60} [45]. Almost simultaneously the synthesis of N@C_{60}O was reported according to Scheme 1 [46].

N@C_{60}/C_{60} was enriched by HPLC to 10^{-3}. The epoxide was formed by reacting **1** with H_2O_2 in the presence of MeO_3Re under ambient conditions for 12 h to give **2**. It was observed that **2** is stable in the dark at room temperature. However, dissolved in toluene and exposed to ambient light, **2** exhibited a linear decay of EPR intensity with a half life of approximately 2 days.

Since that work was published the functionalisation chemistry of N@C_{60} has been expanded. We now know that most additions to the cage inflict some degree of EPR signal loss on N@C_{60}. This implies that either some N@C_{60} is destroyed or that the nitrogen atom escapes from the fullerene cage. The limited availability of high purity N@C_{60} combined with its thermal and photo-instability might initially look like an insurmountable obstacle to further chemistry. However, it is possible to tune reaction conditions in such a manner that a significant "number of spins" survive the reaction. N@C_{60} was reacted with 4-nitrobenzaldehyde and *N*-methylglycine, according to the Prato reaction as shown in Scheme 2 [47].

An excess of N@C_{60} was refluxed with the aldehyde and sarcosine in toluene for 2 h under nitrogen. Under these conditions **3** (N = $C_{69}H_{10}N_2O_2$) was produced with 31% yield. Crutially, the EPR signal intensity of the product mixture was found to be arround 73% of the initial N@C_{60} signal.

One step further is the production of a *half-filled* endohedral fullerene dimer using again the pyrrolidine functionalisation (Scheme 3) [48].

After reflux in toluene for 2 days, the C_{60}-azo-C_{60} dimer was produced in 95% yield. In order to preserve the nitrogen spin signal, the same reaction with N@C_{60} was performed for approximaely 2 h. **4** (N = C_{60}-azo-C_{60}) was produced

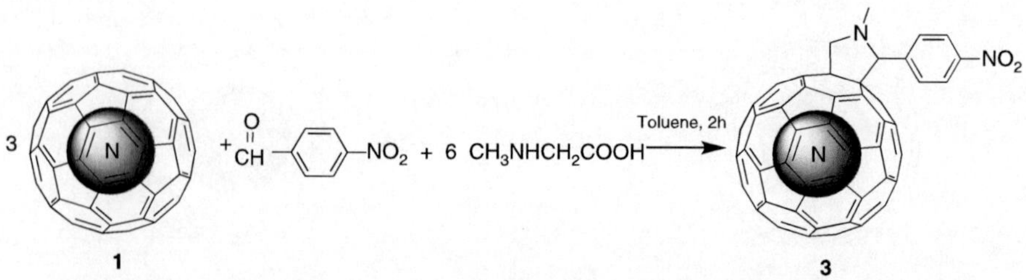

Scheme 2: Synthesis of $N = C_{69}H_{10}N_2O_2$ via the Prato reaction scheme

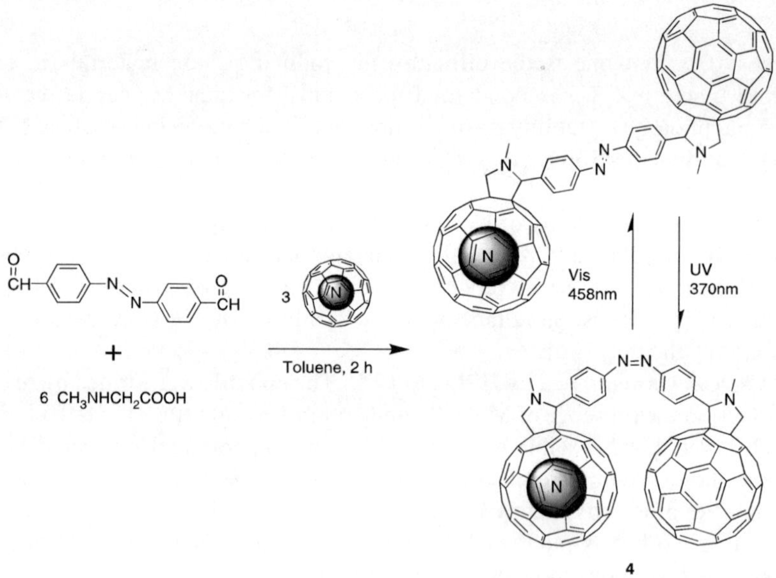

Scheme 3: Synthesis of photo-switchable dimer N@C_{60}-azo-C_{60} and its photoisomerisation

with 30% yield. Its EPR signal was found to be approximately 70% of the starting material. The azobenzene moiety is one of the most efficient photoisomerised molecules. Exposed to visible light it is mostly in the trans- form. Upon irradiation with UV light, it changes to the cis-isomer. We performed the same isomerisation with **4**. Using pulse EPR, molecular rotation correlation times $\tau_c = 37.2 \pm 1.6$ ps for the trans- and $\tau_c = 34.8 \pm 2.7$ ps for the cis-isomers were measured, respectively. This subtle difference is attributed to the difference in size between the two isomers. The trans is the bulkier of the two hence its tumbling is slightly slower than the cis which leads to a longer rotation correlation time.

The beauty of this scheme is that it not only retains most of the N@C_{60} signal, but also affords both the dimer and the monomer products by manipulation of the reagent molar ratios. This is the first time that a two-step reaction to yield an asymmetric endohedral fullerene dimer in a controlled way seems feasible. For example, a $^{14}N@C_{60}$-$^{15}N@C_{60}$ dimer could be possible by first doing the reaction with excess aldehyde to produce the $^{14}N@C_{60}$ monomer and then reacting the monomer with $^{15}N@C_{60}$ to produce the asymmetric dimer. Such a molecule would be the first fullerene two-qubit system. Also the bridge molecule would act s a photo-switch modulating the distance between the fullerene cages and hence the interaction between the qubits. It must be noted that even though this is a unique molecule, a directly bonded $N@C_{60}$-C_{60} dimer was synthesized previously via the high speed vibration milling technique (HSVM) method [49]. However the purity of $N@C_{60}$ was too low to allow for in-depth spectroscopic study of that molecule.

The synthetic schemes analysed in the previous paragraphs involved C_{60} or $N@C_{60}$ the chemistry of which is very similar to C_{60}. Endohedral metallofullerenes have not been used extensively in functionalisation schemes. One reason is that most of them involve C_{82} or C_{80}. These are cages with lower symmetry that give rise to a number of isomers and multi-adducts. On the other hand, metallofullerenes are thermally robust unlike $N@C_{60}$ as we have seen above. Although their chemistry is not so well developed (compared to C_{60}), they are beginning to be available in multi-milligram quantities of high purity materials. Some synthetic protocols for metallofullerene adducts have begun to appear in recent years [50–53]. As more endohedral metallofullerenes become available in larger amounts, their chemistry will be developed further, and sophisticated synthons such as an endohedral metallofullerene dimer may become attainable.

In addition to covalent bonding, non-covalent interactions present an attractive route toward the assembly of arrays of endohedral fullerenes. Such interactions include hydrogen bonding, van der Waals interactions, $\pi - \pi$ stacking interactions and coordination chemistry. Porphyrins, cyclodextrins, calixarenes and other macrocycles can be complexed with fullerenes in order to create supramolecular arrays. Although weak in comparison to covalent bonds, it is well known that very stable structures can be achieved through the cooperative effect of such interactions. An advantage of these interactions is that they are driven with an inherent ability to "self-correct" due to their thermodynamic nature [54]. It looks likely that these schemes can be extended to endohedral fullerenes too.

2.2.8 One-Dimensional, Two-Dimensional Arrays and Beyond

The structure of one-dimensional arrays of fullerenes is relatively straightforward. Fullerene molecules self-assemble into ordered arrays inside SWNTs. The process is spontaneous upon heating and the resulting structures are called nanotube "peapods" [55]. Metallofullerene peapod structures are well established [56, 57]. What is less understood are the peapod electronic structure and the effect of filling on the spin properties of the metallofullerenes. This is partly due to the fact that SWNTs come as a mixture of semi-conducting and metallic ones and with many paramagnetic impurities that interfere with the magnetic properties of the encapsulated endohedral fullerenes. Thermally unstable molecules such as N@C_{60} and its derivatives can also be inserted into SWNTs using solution methods or supercritical fluids [58, 59]. It has even been suggested that the thermal stability of N@C_{60} is enhanced in the peapod structures compared to its crystalline form [60].

Two-dimensional supramolecular structures can form on surfaces by exploiting non-covalent (mainly hydrogen bonding) interactions between the constituent molecules. Some molecular networks can form porous structures that can act as hosts for fullerenes. The arrangement of the guest fullerene molecules is largely controlled by the size and shape of the network pores. Hexagonally packed C_{60} heptamers have been formed in a perylene tetra-carboxylic di-imide (PTCDI)-melamine network on a silver-terminated silicon surface [61]. Single C_{60} molecules have been incorporated in a trimesic acid (TMA) molecular network on graphite [62]. Recently, a strontium titanate ($SrTiO_3$) "waffle" surface was used as a template for arrays of paired endohedral fullerenes ($Er_3N = C_{80}$) [63]. The molecules "fit" like eggs fitting into an egg carton. This work shows that it is possible to arrange endohedral fullerenes in ordered, two-dimensional arrays in a controlled manner and should pave the way for more intricate arrangements of endohedral molecules. In principle, such patterns can be extended in to three-dimensional networks too. Such architectures are prerequisites for many technological applications including nanoelectronics and quantum information processing.

2.3 Graphene

Graphene is a two-dimensional sheet of carbon atoms arranged in a hexagonal structure with sp^2 bonding, shown in Fig. 2.6. Layers of graphene are stacked on top of each other in an AB orientation to form the well- known three-dimensional solid graphite. The layers of graphene are not directly bonded to each but instead are held together within graphite by van der Waals forces. This enables the exfoliation of graphite to form single graphene layers. Two layers of graphene are known as bilayer graphenes, whist 3–6 layers are known as few layer graphenes (Fig. 2.7). Rich physics has been observed in graphene layers [64], along with superior mechanical properties [65]. This has stimulated a buzz in the international community about the possible future exploitation of graphene in electronics and mechanical strengthening applications.

2.3.1 Synthesis

Graphenewas first discovered in 2004 by the Geim group at the University of Manchester [66]. This was the major breakthrough that stimulated an entirely new field. Their process was to mechanically exfoliate graphite by repeatedly

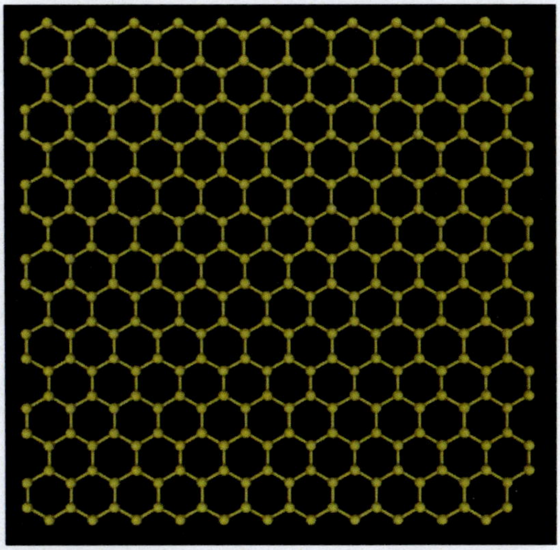

Fig. 2.6 Graphene sheet with arm-chair edge termination at the sides and zigzag termination at top and bottom edges

removing layers of high quality graphite (such as highly orientated pyrolytic graphite (HOPG)) using scotch tape. This reduced the average number of layers in the graphite flakes, which were then stamped onto a silicon substrate coated with a thin 300 nm layer of SiO_2. This process produced only a small number of graphene layers with respect to thicker multi-layered graphite pieces on the surface of the silicon substrate and thus finding graphene was like looking for a needle in a haystack. The key innovation by the Geim group was to use optical microscopy to visually inspect the graphene layers and use a change in the optical contrast that arises from an interference effect caused by the presence of the SiO_2 layer. Optical microscopy enabled thin graphene monolayers to be identified by their weak contrast and then checked with atomic force microscopy to determine the height. Andre Geim and Konstantin Novoselov were awarded the Nobel Prize in Physics 2010 for their groundbreaking experiments with graphene.

While the scotch tape exfoliation method sparked the graphene boom, it is widely acknowledged that this approach has severe limitations in the commercial uptake of graphene in the electronics industry due to the relatively small size of the graphene layers that are produced. Solution phase chemical exfoliation is another technique that has successfully been used to obtain graphene and few layer graphene [67]. In this approach, graphene layers are peeled apart from one another by the combination of ultrasound sonication in a suitable polar solvent or aqueous surfactant system. This leads to flakes of graphene, bilayer graphene and few layer graphene mixed among large chunks of graphite. The large pieces of graphite are removed from the solution by centrifuging at low speeds. Most success in solution phase exfoliation has been achieved using solvents like *N*-methyl-2-pyrrolidone (NMP), Dimethylformamide (DMF), 1,2-dichloroethane and aqueous surfactant systems such as sodium dodecyl sulphate (SDS) and sodium cholate. It seems that currently the chemical exfoliation methods do not produce large area flakes of

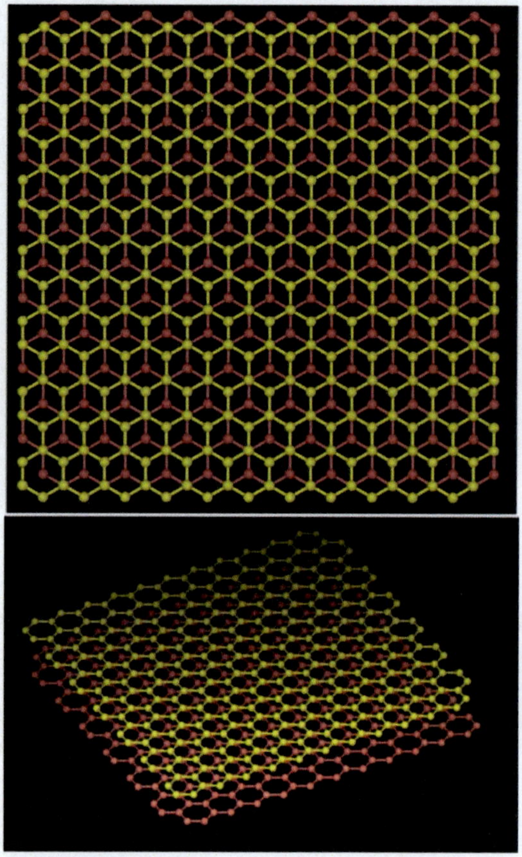

Fig. 2.7 Bilayer graphene with AB stacking. Top layer is *yellow* and bottom layer is *pink*

graphene needed for electronic applications, but can produce macroscopic amounts. The most promising routes for growing large area graphene layers are the surface precipitation growth on SiC substrate [68], and the epitaxial growth on catalytic metals such as nickel [69] and copper [70] using chemical vapour deposition. Graphene grown on these substrates can be transferred to other substrates such as quartz using a variety of techniques [71–73]. The major challenge remaining in this approach is to get the charge carrier mobility up to those reported for the mechanical exfoliation technique.

2.3.2 *Properties and Applications*

Graphene has raised massive interest in the international electronics community by exhibiting ambi-polar electric field effect behaviour at high carrier concentrations of up to 10^{13} cm^{-2} and charge carrier mobilities exceeding 15,000 cm^{2}V^{-1}s^{-1}.

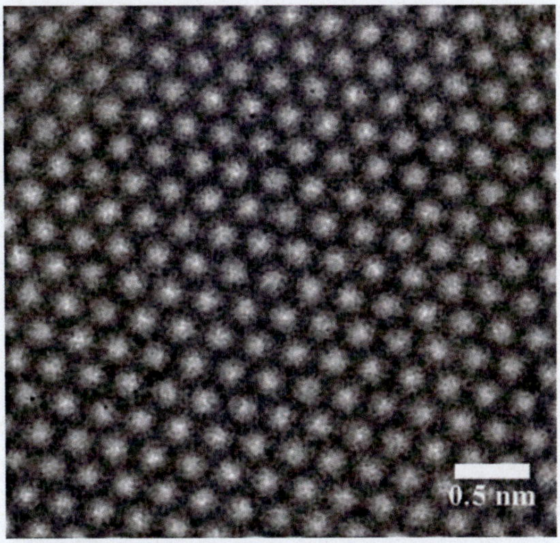

Fig. 2.8 Aberration corrected high-resolution transmission electron microscopy image of graphene taken at an accelerating voltage of 80 kV

Graphene exhibits ballistic charge transport on the sub-micrometer scale [74] and also quantum Hall effects at room temperature [75]. The charge carriers move so fast in graphene that they are best described by the Dirac equation for relativistic particles than the Schrödinger equation. Graphene has immense potential in nanoelectronic applications and for this to be realised, large area mm^2 sheets of high quality graphene or bilayer graphene are needed.

Graphene also boasts one of the highest mechanical strengths ever reported [65]. This makes it very appealing for blending with polymers for reinforcement. Since graphene is only one atomic layer thick and mechanically stable it is ideal for using as an ultrathin transparent support for high resolution electron microscopy [76]. Figure 2.8 shows a high-resolution transmission electron microscopy image of graphene showing the atomic structure.

This image was taken with aberration correction to achieve sub-Angstrom spatial resolution and at a low accelerating voltage of 80 kV in order to minimise damage [77]. The periodic atomic structure is easily removed by post-processing digital filtering leaving images of nanoparticles or molecules as if they were suspended in free space.

2.4 Carbon Nanotubes

A SWNT can be formed by rolling up a graphene sheet to form a tubular structure. The properties of carbon nanotubes are defined by their chirality, which is linked to the way they are rolled up into a cylinder. A chiral vector C_h is used to classify the chirality of a nanotube in terms of (n,m) values, where $C_h = na_1 + ma_2$ as shown in Fig. 2.9. A translational vector T, perpendicular to C, defines the length of the

Fig. 2.9 Determining chirality of single-walled carbon nanotubes

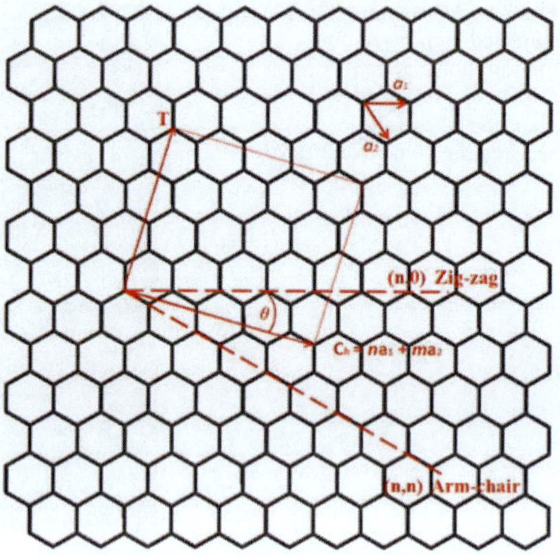

nanotubes' unit cell. The chiral angle θ defines the direction of C relative to the basis vector a_1. The (n,m) values of a nanotube indicate whether it will be semiconducting or metallic. When n-m is divisible by three the nanotubes are typically metallic, alternatively they are semiconducting. Arm-chair SWNTs are defined as $n = m$, i.e., (n,n), zigzag as (n,0), and chiral when $n \neq m$.

As the radius of a nanotubes decreases, the degree of curvature increases. This places significant strain on the sp^2 carbon bonds and leads to an increase in the chemical reactivity of smaller diameter nanotubes and a reduced stability [78], [79]. For semiconducting carbon nanotubes, the bandgap generally increases for decreasing nanotubes radius. This leads to diameter-dependent photoluminescence from SWNTs [80]. Multi-walled carbon nanotubes are formed by concentric addition of larger tubes around a core inner SWNT. Inter-wall coupling leads to MWNTs being primarily metallic. Figure 2.10 shows an end-on perspective view of a (18,0) SWNT.

2.4.1 Synthesis

Observations of nanotubes were reported as far back as 1952 by L. V. Radushkevich and V. M. Lukyanovich in the Soviet *Journal of Physical Chemistry* [81]. However, it is widely acknowledged that the spark that ignited the field of carbon nanotubes was the seminal paper by Sumio Iijima in 1991 [82]. His combination of observation and interpretation was critical in developing nanotube science. The use of transmission electron microscopy was crucial to this discovery. Advances in electron microscopy now enable the atomic structure of nanotubes to be directly imaged, such as in Fig. 2.11 [79].

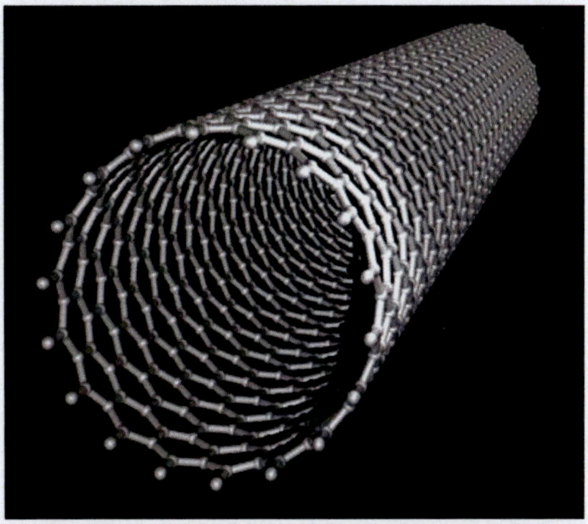

Fig. 2.10 An end-on perspective view of a (18,0) SWNT

Fig. 2.11 Aberration-corrected high-resolution transmission electron microscopy image of a single-walled carbon nanotube showing the atomic structure

Several techniques have been developed to produce carbon nanotubes in large scale quantities. These are arc-discharge [83], laser ablation [84, 85], chemical vapour deposition [86] and high pressure carbon monoxide (HiPCO). Each of these techniques has its own merits depending on what applications the nanotubes are to be used for. Arc-discharge tends to produce large quantities, but often contains a significant amount of catalyst nanoparticles that are usually magnetic and amorphous carbon. Arc-discharge is a desirable route to fabricating nanotubes with diameters between 1.2 and 1.6 nm. Laser ablation tends to produce a higher yield of nanotubes compared to catalysts and amorphous carbon, but does not have the ability to produce the same volume of material that arc-discharge can. HiPCO nanotubes are generally smaller in diameter than arc-discharge or laser ablation with diameters between 0.7 and 1.2 nm. CVD-grown nanotubes often lack the structural quality of nanotubes produced via laser ablation or arc-discharge, but are ideal for growing vertical forests or nanotubes on substrates for nanoelectronics. The diameter distribution of CVD-grown nanotubes is often broader than other methods.

Purification of carbon nanotubes is one of the most challenging aspects of their research. It is important to remove catalyst particles and amorphous carbon to obtain pristine nanotubes. Amorphous carbon is generally removed by either burning in air below the nanotubes decomposition temperature or refluxing in H_2O_2. Catalysts are generally Ni, Co or Fe and can be removed by refluxing in acids such as HCl and HNO_3. Both of these methods also damage the nanotubes and reduce the total amount of product. Alternative way to remove catalyst particles is to use high speed centrifugation methods or to anneal at high temperatures (1,200°C) under high dynamic vacuum. Density gradient ultracentrifugation has recently been demonstrated as a technique to separate metallic and semiconducting nanotubes from each other [87].

2.4.2 Applications

Carbon nanotubes are promising for a wide range of applications involving electronics, mechanical strengthening, photonics and heat dissipation. Semiconducting SWNTs emit photoluminescence in the NIR wavelength region between 1.1 and 2.0 μm, which makes them appealing as biological chromophores and in telecommunications applications. Metallic nanotubes have low resitivity and a high capacity to carry larger currents than metals. The electronic transport of nanotubes shows quantised transport down to the single electron level. Significant advances have been made in developing field effect transistors comprising semiconducting nanotubes operating at GHz frequencies. Young's modulus measurements of nanotubes are on the order of TPa and the incorporation into polymers leads to light-weight ultra-strong composites that are already commercially available. Carbon nanotubes have excellent thermal conductivity up to 2,000 W/m/K, which makes them appealing for heat dissipation. Because of these properties, carbon nanotubes have been hailed as the "wonder material" for the twenty-first century.

2.5 Summary

In this chapter we focused on carbon nanomaterials. We described their synthesis and we analysed their electronic and physico-chemical properties with practical applications in mind. Perhaps the most exotic application is the use of these molecules as building blocks for a quantum device. Quantum phenomena are inherent in atoms and molecules and carbon nanomaterials are the perfect playground to study quantum phenomena even at room temperature. We discussed the synthetic developments in endohedral fullerene chemistry. We focused particularly on various chemical syntheses that have been applied to give endohedral fullerene adducts: from monomers and small fullerene dimers to larger one-dimensional and two-dimensional array architectures.

We discussed the properties of graphene and carbon nanotubes. We showed that quantum phenomena can be studied using these materials and innovative science can be performed at the nanoscale. The unique properties of these materials put them at the forefront of many technologies.

There is little doubt that there are advantages but also complications with carbon nanomaterials and with their potential in applications ranging from medicine to quantum information. It was our intention to highlight the properties that make these molecules unique and the main challenges that need to be overcome. Two Nobel Prizes have been awarded so far for discoveries related to fullerenes and graphene. This is testament of the importance of carbon nanomaterials. Research to date has demonstrated that fullerenes, graphene, carbon nanotubes and their derivatives are not far from finding their way in the market place via established nanotechnological applications.

References

1. E. Osawa, Kagaku (Kyoto) **25**, 854 (1970)
2. H. Kroto, J. Heath, S. O'Brien, R. Curl, R. Smalley, Nature **318**(6042), 162 (1985)
3. W. Kratschmer, L. Lamb, K. Fostiropoulos, D. Huffman, Nature **347**(6291), 354 (1990)
4. K. Kadish, R. Ruoff, *Fullerenes: Chemistry, Physics and Technology* (Wiley, New York, USA, 2000)
5. H. Shinohara, Rep. Prog. Phys. **63**(6), 843 (2000)
6. S. Stevenson, G. Rice, T. Glass, K. Harich, F. Cromer, M. Jordan, J. Craft, E. Hadju, R. Bible, M. Olmstead, K. Maitra, A. Fisher, A. Balch, H. Dorn, Nature **401**(6748), 55 (1999)
7. S. Yang, L. Dunsch, J. Phys. Chem. B **109**(25), 12320 (2005)
8. S. Yang, M. Kalbac, A. Popov, L. Dunsch, Chem. – A Eur. J. **12**(30), 7856 (2006). DOI 10.1002/chem.200600261
9. R. Macfarlane, D. Bethune, S. Stevenson, H. Dorn, Chem. Phys. Lett. **343**(3-4), 29 (2001)
10. S. Stevenson, P. Fowler, T. Heine, J. Duchamp, G. Rice, T. Glass, K. Harich, E. Hajdu, R. Bible, H. Dorn, Nature **408**(6811), 427 (2000)
11. L. Dunsch, S. Yang, Small **3**(8), 1298 (2007). DOI 10.1002/smll.200700036
12. T. Murphy, T. Pawlik, A. Weidinger, M. Hohne, R. Alcala, J. Spaeth, Phys. Rev. Lett. **77**(6), 1075 (1996)
13. A. Weidinger, M. Waiblinger, B. Pietzak, T. Murphy, Appl. Phys. A – Mater. Sci. Process. **66**(3), 287 (1998)
14. T. Kaneko, S. Abe, H. Ishida, R. Hatakeyama, Phys. Plasmas **14**(11) (2007). DOI 10.1063/1.2814049
15. K. Komatsu, M. Murata, Y. Murata, Science **307**(5707), 238 (2005). DOI 10.1126/science.1106185
16. Y. Murata, M. Murata, K. Komatsu, J. Am. Chem. Soc. **125**(24), 7152 (2003). DOI 10.1021/ja0354162
17. H. Okimoto, R. Kitaura, T. Nakamura, Y. Ito, Y. Kitamura, T. Akachi, D. Ogawa, N. Imazu, Y. Kato, Y. Asada, T. Sugai, H. Osawa, T. Matsushita, T. Muro, H. Shinohara, J. Phys. Chem. C **112**(15), 6103 (2008). DOI 10.1021/jp711776j
18. T. Akasaka, S. Okubo, M. Kondo, Y. Maeda, T. Wakahara, T. Kato, T. Suzuki, K. Yamamoto, K. Kobayashi, S. Nagase, Chem. Phys. Lett. **319**(1–2), 153 (2000)
19. D. Leigh, J. Owen, S. Lee, K. Porfyrakis, A. Ardavan, T. Dennis, D. Pettifor, G. Briggs, Chem. Phys. Lett. **414**(4–6), 307 (2005). DOI 10.1016/j.cplett.2005.08.090
20. T. Suetsuna, N. Dragoe, W. Harneit, A. Weidinger, H. Shimotani, S. Ito, H. Takagi, K. Kitazawa, Chem. – A Eur. J. **8**(22), 5079 (2002)
21. P. Jakes, K. Dinse, C. Meyer, W. Harneit, A. Weidinger, Phys. Chem. Chem. Phys. **5**(19), 4080 (2003)

22. M. Kanai, K. Porfyrakis, A. Briggs, T. Dennis, Chem. Commun. (2), 210 (2004). DOI 10.1039/b310978h
23. A. Bartl, L. Dunsch, U. Kirbach, G. Seifert, Synth. Met. **86**(1–3), 2395 (1997). International Conference on the Science and Technology of Synthetic Metals, Snowbird, UT, July 28–August 02, 1996
24. M.A.G. Jones, R.A. Taylor, A. Ardavan, K. Porfyrakis, G.A.D. Briggs, Chem. Phys. Lett. **428**(4–6), 303 (2006). DOI 10.1016/j.cplett.2006.06.094
25. E. Xenogiannopoulou, S. Couris, E. Koudoumas, N. Tagmatarchis, T. Inoue, H. Shinohara, Chem. Phys. Lett. **394**(1–3), 14 (2004). DOI 10.1016/j.cplett.2004.06.093
26. R. Bolskar, A. Benedetto, L. Husebo, R. Price, E. Jackson, S. Wallace, L. Wilson, J. Alford, J. Amer. Chem. Soc. **125**(18), 5471 (2003). DOI 10.1021/ja03400984
27. D. Cagle, S. Kennel, S. Mirzadeh, J. Alford, L. Wilson, Proc. Natl. Acad. Sci. USA **96**(9), 5182 (1999)
28. J.J. Yin, F. Lao, J. Meng, P.P. Fu, Y. Zhao, G. Xing, X. Gao, B. Sun, P.C. Wang, C. Chen, X.J. Liang, Mol. Pharmacol. **74**(4), 1132 (2008). DOI 10.1124/mol.108.048348
29. W. Ma, C. Yang, X. Gong, K. Lee, A. Heeger, Adv. Funct. Mater. **15**(10), 1617 (2005). DOI 10.1002/adfm.200500211
30. R.B. Ross, C.M. Cardona, D.M. Guldi, S.G. Sankaranarayanan, M.O. Reese, N. Kopidakis, J. Peet, B. Walker, G.C. Bazan, E. Van Keuren, B.C. Holloway, M. Drees, Nat. Mater. **8**(3), 208 (2009). DOI 10.1038/NMAT2379
31. G. Morley, B. Herbert, S. Lee, K. Porfyrakis, T. Dennis, D. Nguyen-Manh, R. Scipioni, J. van Tol, A. Horsfield, A. Ardavan, D. Pettifor, J. Green, G. Briggs, Nanotechnology **16**(11), 2469 (2005). DOI 10.1088/0957-4484/16/11/001
32. B. Plakhutin, N. Breslavskaya, E. Gorelik, A. Arbuznikov, J. Mol. Struct.-Theorem **727**(1–3), 149 (2005). DOI 10.1016/j.theochem.2005.02.031
33. H. Mauser, N. Hommes, T. Clark, A. Hirsch, B. Pietzak, A. Weidinger, L. Dunsch, Angew Chem Int Ed **36**(24), 2835 (1997)
34. A. Steane, Phys. Rev. Lett. **77**(5), 793 (1996)
35. L. Vandersypen, M. Steffen, G. Breyta, C. Yannoni, M. Sherwood, I. Chuang, Nature **414**(6866), 883 (2001)
36. D. DiVincenzo, Fortschritte der physik-progress of Physics **48**(9–11), 771 (2000)
37. C. Bennett, D. DiVincenzo, Nature **404**(6775), 247 (2000)
38. J. Morton, A. Tyryshkin, A. Ardavan, K. Porfyrakis, S. Lyon, G. Briggs, J. Chem. Phys. **124**(1) (2006). DOI 10.1063/1.2147262
39. W. Harneit, Phys. Rev. A **65**(3, Part A), 032322 (2002). DOI 10.1103/PhysRevA.65
40. A. Ardavan, M. Austwick, S. Benjamin, G. Briggs, T. Dennis, A. Ferguson, D. Hasko, M. Kanai, A. Khlobystov, B. Lovett, G. Morley, R. Oliver, D. Pettifor, K. Porfyrakis, J. Reina, J. Rice, J. Smith, R. Taylor, D. Williams, C. Adelmann, H. Mariette, R. Hamers, Philos. Trans. R. Soc. Lond. Ser. A – Math. Phys. Eng. Sci. **361**(1808), 1473 (2003). DOI 10.1098/rsta.2003.1214. Discussion Meeting of the Royal-Society, London, England, March 13–14, 2002
41. S. Benjamin, A. Ardavan, G. Andrew, D. Briggs, D. Britz, D. Gunlycke, J. Jefferson, M. Jones, D. Leigh, B. Lovett, A. Khlobystov, S. Lyon, J. Morton, K. Porfyrakis, M. Sambrook, A. Tyryshkin, J. Phys. Condens. Matter **18**(21, Sp. Iss. SI), S867 (2006). DOI 10.1088/0953-8984/18/21/S12
42. R. Taylor, *Lecture Notes on Fullerene Chemistry: A Handbook for Chemists* (Imperial College Press, London, UK, 1999)
43. A. Hirsch, M. Brettreich, F.B. Wudl, *Fullerenes: Chemistry and Reactions* (Wiley VCH, Weinheim, Germany, 2004)
44. B. Pietzak, M. Waiblinger, T. Murphy, A. Weidinger, M. Hohne, E. Dietel, A. Hirsch, Chem. Phys. Lett. **279**(5–6), 259 (1997)
45. L. Franco, S. Ceola, C. Corvaja, S. Bolzonella, W. Harneit, M. Maggini, Chem. Phys. Lett. **422**(1–3), 100 (2006). DOI 10.1016/j.cplett.2006.02.046
46. M. Jones, D. Britz, J. Morton, A. Khlobystov, K. Porfyrakis, A. Ardavan, G. Briggs, Phys. Chem. Chem. Phys. **8**(17), 2083 (2006). DOI 10.1039/b601171c

47. J. Zhang, J.J.L. Morton, M.R. Sambrook, K. Porfyrakis, A. Ardavan, G.A.D. Briggs, Chem. Phys. Lett. **432**(4–6), 523 (2006). DOI 10.1016/j.cplett.2006.10.116
48. J. Zhang, K. Porfyrakis, J.J.L. Morton, M.R. Sambrook, J. Harmer, L. Xiao, A. Ardavan, G.A.D. Briggs, J. Phys. Chem. C **112**(8), 2802 (2008). DOI 10.1021/jp711861z
49. B. Goedde, M. Waiblinger, P. Jakes, N. Weiden, K. Dinse, A. Weidinger, Chem. Phys. Lett. **334**(1–3), 12 (2001)
50. T. Wakahara, Y. Iiduka, O. Ikenaga, T. Nakahodo, A. Sakuraba, T. Tsuchiya, Y. Maeda, M. Kako, T. Akasaka, K. Yoza, E. Horn, N. Mizorogi, S. Nagase, J. Amer. Chem. Soc. **128**(30), 9919 (2006). DOI 10.1021/ja062233h
51. O. Lukoyanova, C.M. Cardona, J. Rivera, L.Z. Lugo-Morales, C.J. Chancellor, M.M. Olmstead, A. Rodriguez-Fortea, J.M. Poblet, A.L. Balch, L. Echegoyen, J. Amer. Chem. Soc. **129**(34), 10423 (2007). DOI 10.1021/ja071733n
52. Y. Takano, A. Yomogida, H. Nikawa, M. Yamada, T. Wakahara, T. Tsuchiya, M.O. Ishitsuka, Y. Maeda, T. Akasaka, T. Kato, Z. Slanina, N. Mizorogi, S. Nagase, J. Amer. Chem. Soc. **130**(48), 16224 (2008). DOI 10.1021/ja802748q
53. T. Akasaka, T. Kono, Y. Takematsu, H. Nikawa, T. Nakahodo, T. Wakahara, M.O. Ishitsuka, T. Tsuchiya, Y. Maeda, M.T.H. Liu, K. Yoza, T. Kato, K. Yamamoto, N. Mizorogi, Z. Slanina, S. Nagase, J. Amer. Chem. Soc. **130**(39), 12840+ (2008). DOI 10.1021/ja802156n
54. J. Lindsey, New J. Chem. **15**(2–3), 153 (1991)
55. B. Smith, M. Monthioux, D. Luzzi, Nature **396**(6709), 323 (1998)
56. K. Hirahara, K. Suenaga, S. Bandow, H. Kato, T. Okazaki, H. Shinohara, S. Iijima, Phys. Rev. Lett. **85**(25), 5384 (2000)
57. J.H. Warner, A.A.R. Watt, L. Ge, K. Porfyrakis, T. Akachi, H. Okimoto, Y. Ito, A. Ardavan, B. Montanari, J.H. Jefferson, N.M. Harrison, H. Shinohara, G.A.D. Briggs, Nano Lett. **8**(4), 1005 (2008). DOI 10.1021/nl0726104
58. F. Simon, H. Kuzmany, H. Rauf, T. Pichler, J. Bernardi, H. Peterlik, L. Korecz, F. Fulop, A. Janossy, Chem. Phys. Lett. **383**(3–4), 362 (2004). DOI 10.1016/j.cplett.2003.11.039
59. A. Khlobystov, D. Britz, J. Wang, S. O'Neil, M. Poliakoff, G. Briggs, J. Mater. Chem. **14**(19), 2852 (2004). DOI 10.1039/b404167d
60. S. Toth, D. Quintavalle, B. Nafradi, L. Korecz, L. Forro, F. Simon, Phys. Rev. B **77**(21) (2008). DOI 10.1103/PhysRevB.77.214409
61. J. Theobald, N. Oxtoby, M. Phillips, N. Champness, P. Beton, Nature **424**(6952), 1029 (2003). DOI 10.1038/nature01915
62. S. Griessl, M. Lackinger, F. Jamitzky, T. Markert, M. Hietschold, W. Heckl, J. Phys. Chem. B **108**(31), 11556 (2004). DOI 10.1021/jp049521p
63. D.S. Deak, F. Silly, K. Porfyrakis, M.R. Castell, J. Amer. Chem. Soc. **128**(43), 13976 (2006). DOI 10.1021/ja0634369
64. A.K. Geim, K.S. Novoselov, Nat. Mater. **6**(3), 183 (2007)
65. C. Lee, X. Wei, J.W. Kysar, J. Hone, Science **321**(5887), 385 (2008). DOI 10.1126/science.1157996
66. K. Novoselov, A. Geim, S. Morozov, D. Jiang, Y. Zhang, S. Dubonos, I. Grigorieva, A. Firsov, Science **306**(5696), 666 (2004)
67. Y. Hernandez, V. Nicolosi, M. Lotya, F.M. Blighe, Z. Sun, S. De, I.T. McGovern, B. Holland, M. Byrne, Y.K. Gun'ko, J.J. Boland, P. Niraj, G. Duesberg, S. Krishnamurthy, R. Goodhue, J. Hutchison, V. Scardaci, A.C. Ferrari, J.N. Coleman, Nat. Nanotechnol. **3**(9), 563 (2008). DOI 10.1038/nnano.2008.215
68. C. Berger, Z. Song, X. Li, X. Wu, N. Brown, C. Naud, D. Mayou, T. Li, J. Hass, A.N. Marchenkov, E.H. Conrad, P.N. First, W.A. de Heer, Science **312**(5777), 1191 (2006). DOI 10.1126/science.1125925
69. A. Reina, X. Jia, J. Ho, D. Nezich, H. Son, V. Bulovic, M.S. Dresselhaus, J. Kong, Nano Lett. **9**(1), 30 (2009). DOI 10.1021/nl801827v
70. X. Li, W. Cai, J. An, S. Kim, J. Nah, D. Yang, R. Piner, A. Velamakanni, I. Jung, E. Tutuc, S.K. Banerjee, L. Colombo, R.S. Ruoff, Science **324**(5932), 1312 (2009). DOI 10.1126/science.1171245

71. K. Novoselov, D. Jiang, F. Schedin, T. Booth, V. Khotkevich, S. Morozov, A. Geim, Proc. Nat. Acad. Sci. USA **102**(30), 10451 (2005). DOI 10.1073/pnas.0502848102
72. K. Novoselov, A. Geim, S. Morozov, D. Jiang, M. Katsnelson, I. Grigorieva, S. Dubonos, A. Firsov, Nature **438**(7065), 197 (2005). DOI 10.1038/nature04233
73. Y. Zhang, Y. Tan, H. Stormer, P. Kim, Nature **438**(7065), 201 (2005). DOI 10.1038/nature04235
74. X. Du, I. Skachko, A. Barker, E.Y. Andrei, Nat. Nanotechnol. **3**(8), 491 (2008). DOI 10.1038/nnano.2008.199
75. K.S. Novoselov, Z. Jiang, Y. Zhang, S.V. Morozov, H.L. Stormer, U. Zeitler, J.C. Maan, G.S. Boebinger, P. Kim, A.K. Geim, Science **315**(5817), 1379 (2007). DOI 10.1126/science.1137201
76. J.C. Meyer, C.O. Girit, M.F. Crommie, A. Zettl, Nature **454**(7202), 319 (2008). DOI 10.1038/nature07094
77. J.H. Warner, M.H. Ruemmeli, L. Ge, T. Gemming, B. Montanari, N.M. Harrison, B. Buechner, G.A.D. Briggs, Nat. Nanotechnol. **4**(8), 500 (2009). DOI 10.1038/NNANO.2009.194
78. S. Niyogi, M. Hamon, H. Hu, B. Zhao, P. Bhowmik, R. Sen, M. Itkis, R. Haddon, Acc. Chem. Res. **35**(12), 1105 (2002). DOI 10.1021/ar010155r
79. J.H. Warner, F. Schaeffel, G. Zhong, M.H. Ruemmeli, B. Buechner, J. Robertson, G.A.D. Briggs, ACS Nano **3**(6), 1557 (2009). DOI 10.1021/nn900362a
80. M. O'Connell, S. Bachilo, C. Huffman, V. Moore, M. Strano, E. Haroz, K. Rialon, P. Boul, W. Noon, C. Kittrell, J. Ma, R. Hauge, R. Weisman, R. Smalley, Science **297**(5581), 593 (2002)
81. L. Radushkevich, V. Lukyanovich, Zurn Fisic Chim **26**, 88 (1952)
82. S. Iijima, Nature **354**(6348), 56 (1991)
83. T. Ebbesen, P. Ajayan, Nature **358**(6383), 220 (1992)
84. T. Guo, P. Nikolaev, A. Rinzler, D. Tomanek, D. Colbert, R. Smalley, J. Phys. Chem. **99**(27), 10694 (1995)
85. T. Guo, P. Nikolaev, A. Thess, D. Colbert, R. Smalley, Chem. Phys. Lett. **243**(1–2), 49 (1995)
86. M. Joseyacaman, M. Mikiyoshida, L. Rendon, J. Santiesteban, Appl. Phys. Lett. **62**(6), 657 (1993)
87. M.S. Arnold, A.A. Green, J.F. Hulvat, S.I. Stupp, M.C. Hersam, Nat. Nanotechnol. **1**(1), 60 (2006). DOI 10.1038/nnano.2006.52

Chapter 3
Carbon Nanotubes: From Symmetry to Applications

M. Damnjanović

Abstract In this chapter, we show how the concept of symmetry gives theoretical explanation of the properties, which made carbon nanotubes (NTs) one of the most interesting materials of nanotechnology. First, in Sect. 3.1, we consider basic facts on single-wall carbon nanotubes (SWCNTs), including their configuration and symmetry. Then, we discuss double-wall nanotubes.

Next, Sect. 3.2 is devoted to elementary symmetry-based physical properties. More precisely, we explain the energy spectrum of electrons and phonons, showing that as the consequence of the symmetry, energies must be arranged in the so-called bands. Elementary properties of these band structures may be a priory discussed, yielding easily famous conducting law, showing strong dependence of conductivity on the type of nanotube. Conserved quantum numbers enable us to extract selection rules for various physical processes. This way, radial breathing mode appears to be very important for the characterization of the samples by Raman spectroscopy. Also, optical properties are derived.

Finally, in Sect. 3.3, mutual interaction between the walls of double-wall nanotubes is discussed. It is explained why this interaction is very weak, which is used to propose nanomachines with almost superslippery parts.

3.1 Introduction: Symmetry of Nanotubes

From the very discovery [1], carbon nanotubes are probably the most interesting objects of material science. Their unique properties are investigated in almost all fields of natural sciences, with applications relating diverse parts of technology [2]. It was immediately clear that these structures are highly symmetric, and this was

M. Damnjanović (✉)
NanoLab, Faculty of Physics, University of Belgrade, POB 368, Belgrade 11001, Serbia
e-mail: yqoq@rcub.bg.ac.rs

S. Logothetidis (ed.), *Nanostructured Materials and Their Applications*,
NanoScience and Technology, DOI 10.1007/978-3-642-22227-6_3,

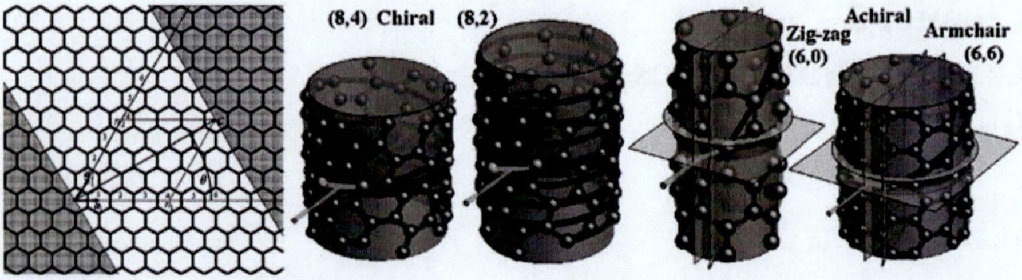

Fig. 3.1 *Left*: graphene, with a chiral vector of (4,4) nanotube (shaded part would be cut out in the rolling). *Right*: nanotubes with depicted helix, U-axis, mirror planes (rectangular), roto-reflectional (round), and glide (zig-zag) planes

used even in the early prediction of their remarkable conducting properties [3–5]. Still, only in 1999 the full symmetry of these structures was found [6] and described by the line groups, which is latter on intensively used in the literature [7, 8]. In this section, a brief reminder of carbon nanotubes configuration will be given, to enable us to determine their symmetry.

Most of the experimentally grown carbon nanotubes are multi-walled, i.e., they have several coaxial single-wall tubes, each of them being a rolled up graphene layer, with the difference of radii of the adjacent walls similar to the distance of 3.44Å of the layers in graphite.

3.1.1 Configuration of Single-Wall Nanotubes

SWCNTs are most easily described as rolled up graphene (Fig. 3.1). Graphene, only recently synthesized [9] honey-comb like layer of graphite, has a symmetry of the symmorphic diperiodic group DG80, with the isogonal point group $\boldsymbol{D}_{6\mathrm{h}}$ and hexagonal lattice with both periods $\boldsymbol{a}_1$ and $\boldsymbol{a}_2$ of equal length of 2.41Å. One gets SWCNT (n_1, n_2) (integers n_1 and n_2 are known as chiral indices) by cutting from the graphene a ribbon perpendicular to the so-called chiral vector $\boldsymbol{c} = n_1\boldsymbol{a}_1 + n_2\boldsymbol{a}_2$, and rolling it in the way that $\boldsymbol{c}$ becomes circumferential to the obtained cylinder. Alternatively, chiral vector is defined by its length (related to the diameter D of the nanotube by $c = D\pi$) and the chiral angle θ.

3.1.2 Symmetry of Single-Wall Nanotubes

Due to the symmetries of graphene layer, in particular rotational axis of order 6 and vertical mirror planes, the different tubes are exhausted by the chiral angle of different tubes in the range $0 - 30°$; precisely, tubes with θ between $60°$ and $30°$ are their optical isomers, usually treated as equivalent. The limiting values

0° and 30° correspond to the achiral nanotubes $(n, 0)$ and (n, n), called zig-zag (Z) and armchair (A); all others are chiral (C). The symmetry of the nanotube (n_1, n_2) comes from the symmetry of graphene [6]: layer translations become roto-helical operations and rotation for π (i.e., C_6^3) of graphene gives horizontal U-axis of the nanotube. Additionally, zig-zig chiral vector is within the layer vertical mirror plane (along $\boldsymbol{a}_1$), which yields horizontal mirror plane of the tube; the perpendicular mirror plane of graphene remains vertical mirror plane of the tube. Analogously, graphene mirror planes with $\boldsymbol{a}_1 + \boldsymbol{a}_2$ and perpendicular to it give horizontal and vertical mirror planes of armchair tubes. Consequently, the symmetry of nanotube is described by line group; depending on the chirality it is

$$\boldsymbol{L}_{\mathcal{C}} = \boldsymbol{T}_q^r(f)\boldsymbol{D}_n = \boldsymbol{L}q_p22, \quad \boldsymbol{L}_{\mathcal{ZA}} = \boldsymbol{T}_{2n}^1(f)\boldsymbol{D}_{nh} = \boldsymbol{L}2n_n/mcm, \tag{3.1a}$$

where the parameters n (order of the principle axis), f (fractional translation), q (one half of the number of the atoms per period), and r and p (helicity parameters) are

$$n = \mathrm{GCD}(n_1, n_2),\ q = 2\frac{n_1^2 + n_1 n_2 + n_2^2}{q\mathcal{R}},\ f = \frac{\sqrt{3(n_1^2 + n_2^2 + n_1 n_2)}}{n\mathcal{R}} a_0, \tag{3.1b}$$

$$r = \frac{n_1 + 2n_2 - (\frac{n_2}{n})^{\phi(\frac{n_1}{n})-1} q\mathcal{R}}{n_1\mathcal{R}} \mod \frac{q}{n},\ p = n\left(r^{\phi(\frac{q}{n})-1} \mod \frac{q}{n}\right) \tag{3.1c}$$

($\mathcal{R} = \mathrm{GCD}(2n_1 + n_2, n_1 + 2n_2)/n$, $a_0 = 2.46$ Å is graphene period and $\phi(x)$ is the Euler function, i.e., number of coprimes not greater then x). Note that the period of the tubes is related to the fractional translation as $a = qf/n$. For achiral tubes $q = 2n$, $r = 1$, and $a_Z = \sqrt{3}a_0$, $a_A = a_0$. The helical group $\boldsymbol{T}_q^r(f)$ is generated by $(C_q^r|f)$, i.e., rotation for $2\pi r/q$ around the tube axis (z-axis) followed by the translation for f along it.

All tubes are with high order helical axis. In fact, q is even, and its form is $q = (12K + 2)n$, with $K = 0, 1, \ldots$; only for achiral tubes, it gets it minimal value $q = 2n$, while for chiral tubes it is a large integer. Different tubes have different roto-helical symmetries, i.e., chiral indices and triples ($Q = q/r, n, f$) are biuniquely related. Even more, all chiral tubes have different pairs (Q, n), while groups of achiral tubes $(n, 0)$ and (n, n) differ only in the fractional translations.

Isogonal point group is $\boldsymbol{D}_q$ for chiral and $\boldsymbol{D}_{2n\mathrm{h}}$ for achiral tubes; obviously, q is the order of its principle axis.

When roto-helical transformations and U-axis act to an arbitrary atom, the whole nanotube is obtained. This means that all single-wall nanotubes are mono-orbit systems (generated by symmetry from a single atom). Therefore, it is intuitively clear that all properties of nanotube are determined by symmetry and single atom.

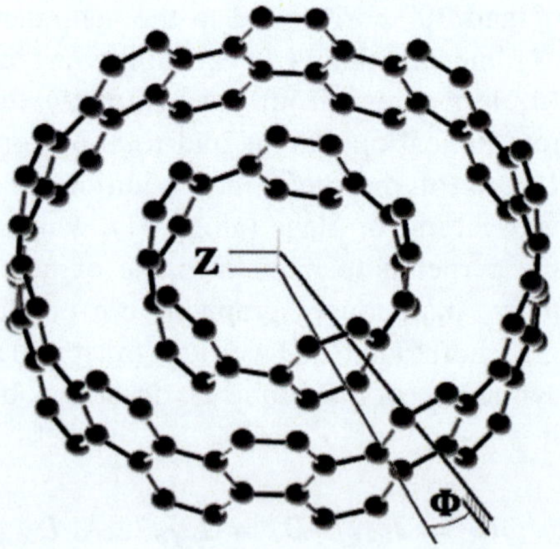

Fig. 3.2 Configuration and relative coordinates of double-wall nanotube

Conveniently, we define nanotube reference frame with z-axis being the nanotube axis, and x-axis coinciding with the U-axis through the center of carbon hexagons. Then, for the initial atom C_{000}, we chose one with cylindrical coordinates:

$$\boldsymbol{r}_{000} = \left(\frac{D}{2}, \varphi_{000} = 2\pi \frac{n_1 + n_2}{nq\mathcal{R}}, z_{000} = \frac{n_1 - n_2}{\sqrt{6nq\mathcal{R}}} a_0\right). \tag{3.2}$$

Since any other atom is obtained by the action of group transformation, we denote atom obtained from C_{000} by $\ell_{tsu} = (C_q^r|\frac{n}{q}a)^t C_n^s U^u$ as C_{tsu}. Its coordinates are:

$$\boldsymbol{r}_{tsu} = \ell_{tsu}\boldsymbol{r}_{000} = \left(\frac{D}{2}, (-1)^u \varphi_{000} + 2\pi\left(\frac{rt}{q} + \frac{s}{n}\right), (-1)^u z_{000} + t\frac{n}{q}a\right). \tag{3.3}$$

3.1.3 Double-Wall Nanotubes

Although we consider the double-wall nanotubes, all the methods and results presented are easily generalized to multi-wall cases. The double-wall nanotube $W@W' = (n_1, n_2)@(n_1', n_2')$ consists of coaxially arranged single-wall ones, with walls denoted as $W = (n_1, n_2)$ and $W' = (n_1', n_2')$. Since the axes z and z' of the walls coincide, their relative position is determined by the angle Φ and vertical displacement Z of their reference frames (Fig. 3.2: x-axis, defined by (3.2), should be rotated by Φ and upraised by Z to get x'-axis of the outer wall. Therefore, in the frame of the interior wall, its atoms have coordinates (3.3), while for the positions $\boldsymbol{r}'_{tsu}$ of the outer wall atoms we use the same expression, but with Φ and Z added to φ'_{000} and z'_{000}, respectively.

Symmetry group of a multi-wall carbon nanotube is the intersection of the symmetry groups (3.1) of the walls. Most of the double-wall tubes are incommensurate, i.e., there is no pure translation (or roto-helical operation, as it can be proved) leaving the whole compound invariant). In fact, although each wall is highly symmetric, their symmetries are mostly incompatible, giving comparatively low symmetry of the total system, which in the incommensurate cases reduces to a point group. This has strong impact to the interaction of the walls.

3.2 Energy Bands

As a consequence of the translational symmetry, SWCNTs have conserved quasi-momentum k: eigenenergies of electrons and phonons, as well as of other (quasi)particle excitations, and are distributed in bands $E(k)$ over Brillouin zone $k \in (\pi/a, \pi/a)$. Moreover, due to the U-axis, there is a band degeneracy $E(k) = E(-k)$, enabling us to consider eigenproblems in the irreducible domain $k \in (0, \pi/a)$ only. The number of these bands depends on the type of tube and (quasi)particles. Different bands are labeled by other conserved quantum numbers (angular momentum m with integral values from $(-q/2, q//2]$, and parities). Only if for several bands all the quantum numbers are equal, we introduce additional label counting them.

3.2.1 Electronic Bands

The simplest tight-binding dynamical model of the electrons is spin independent sp^2 model: carbon $1s$ orbital is occupied by the two localized core electrons, while the bonds with the nearest neighbors are realized by three bonding hybridized orbitals ($2s$ and two in-plane $2p$), occupied by three electrons per atom. Therefore, the relevant (for dynamics) electronic state space is spanned by the remaining p-orbital $|\, tsu\rangle$ from each atom, which is perpendicular to the tube surface. The remaining electron per atom makes it half-filled. The orbitals from the different sites are assumed orthogonal. Hamiltonian is built within the nearest neighbors approximation, with three symmetrically distributed neighbors. Solving its eigenproblem, the energy bands are obtained:

$$\epsilon_m^{\pm}(k) = E_{\mathrm{C}} \pm |V| \sqrt{\sum_{i=1}^{3}(1 + 2\cos\psi_i)}. \tag{3.4}$$

Here, angles ψ_i characterize positions of the three first neighbors to the initial atom C_{000}, while E_{C} is carbon $2p$ orbital eigenenergy.

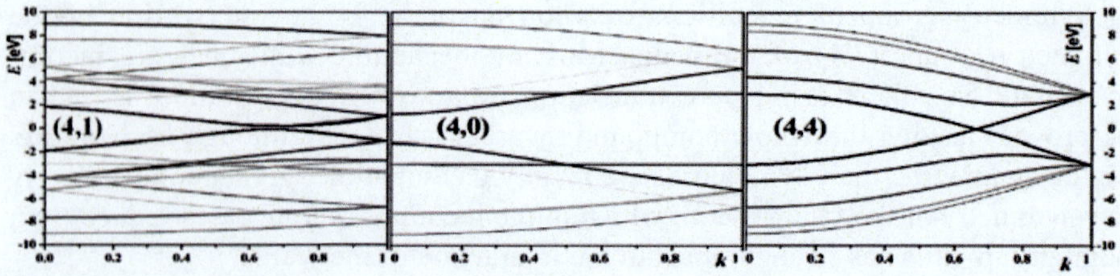

Fig. 3.3 Electronic bands of a chiral (quasimetalic), zig-zag (semiconducting), and armchair (metalic) nanotubes within the simplest p-orbital, nearest neighbor tight-binding approximation

3.2.1.1 Conductivity

Obviously, for each m there are two bands $\epsilon_m^+(k)$ and $\epsilon_m^-(k)$, symmetric with respect to E_C (Fig. 3.3). Taking into account that only half of the states are occupied (at zero temperature), we see that Fermi level E_F coincides with E_C (therefore, according to the convention $E_F = 0$, it is usual to take E_C in (3.4)). The bands $\epsilon_m^+(k)$ are conducting, while valence ones are $\epsilon_m^-(k)$. Consequently, there is a gap between conducting and valence bands (and the tube is semiconducting) unless for some m and k the square root in (3.4) vanishes. Simple calculation shows that this occurs only for the tubes with $n_1 - n_2$ divisible by 3, i.e., only these tubes are conducting.

However, Landaus's non-crossing rule [10] for the bands infers more subtle symmetry based detail. Indeed, having all quantum numbers the same, the bands $\epsilon_m^+(k)$ and $\epsilon_m^-(k)$ cannot cross, i.e., in a more precise model small secondary gap appears. Still, for armchair tubes this occurs (as shown on the right of Fig. 3.3), because the crossing bands have $m = 0$: then there are two representations of the opposite vertical mirror plane parity, and it is easily checked that just the bands with different parity are crossed.

Precise calculations, verified by experimental data, show that this picture is quite well, except for the ultra narrow tubes (D less than 4 Å). In fact, due to strong curvature effects, all these tubes are conducting (note that only few tubes, including (4,0), belong to this class).

To summarize, there are three types of nanotubes: ultra thin and armchair ones are conducting, the remaining tubes with $n_1 - n_2$ divisible by three are quasiconducting (due to a small gap of 0.01 eV, they are conducting on the room temperature), and all others are semiconducting (with gaps of order of 1 eV). Calculations show that gap decreases with the diameter of the tubes, vanishing in the infinite limit, when graphene, being a zero-gap semiconductor, is obtained. Just for this reason nanotubes are interesting in nanoelectronics, offering conducting properties as needed in each particular nanodevice.

3.2.1.2 Optical Properties

Another possible application of nanotubes related to electronic bands is based on their optical absorption. It turns out that depending on the polarization of the

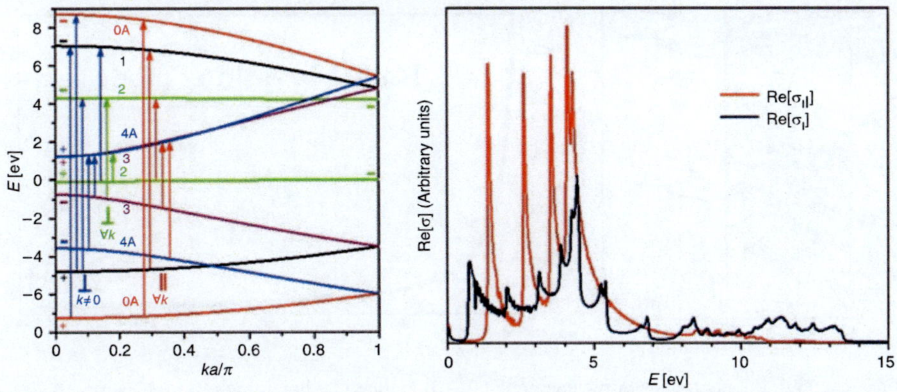

Fig. 3.4 Optical absorption in nanotube (4,0). *Left*: optical transitions allowed by the selection rules (3.5); polarization of the electric field is denoted by ∥ and ⊥. *Right*: dependence of absorption intensities on the energy of the light for two polarizations

incoming light, different frequencies are absorbed. This is due to the selection rules for the electronic transitions induced by excitations by absorbed photons. Namely, these rules differ when the electric field is along and perpendicular to the tube:

$$\parallel : \Delta k = 0,\ \Delta m = 0,\ \Pi_{U\mathrm{f}}\Pi_{U\mathrm{i}} \neq 1,\ \Pi_{\mathrm{vf}}\Pi_{\mathrm{vi}} \neq -1,\ \Pi_{\mathrm{hf}}\Pi_{\mathrm{hi}} \neq 1; \quad (3.5a)$$

$$\perp : \quad \Delta k = 0,\ \Delta m = \pm 1,\ \Pi_{\mathrm{hf}}\Pi_{\mathrm{hi}} \neq -1; \quad (3.5b)$$

here, subscripts i and f denote initial and final electronic state. Consequently, different transitions are allowed for two polarizations of incoming light. This is illustrated in Fig. 3.4.

3.2.2 *Phonons*

Analogously to the electronic energies, also vibrational ones are structured in bands, which are along the irreducible domain assigned by the quantum numbers of angular momentum and parities. Avoiding here the general discussion on phonons, we focus on the Raman active ones. Raman scattering (in its somewhat simplified model) is a process in which one electron is excited by absorbing light, and then relax in two steps, firstly exciting ions (i.e., transmitting a part of energy to phonons) and then recombining to the initial state by light emission. Using the selection rules for each of these three steps, one finds that the selection rules admit excitation of the phonons with quantum number $k = 0$. In addition, for chiral tubes $m = 0, 1$ (irrespectively on U-parity); for achiral tubes active are $m = 0, 2$ with σ_{h} even, and $m = 1$ with σ_{h} odd phonons. Consequently, the number of the Raman active modes is 26, 14, and 15 for chiral, zig-zag, and armchair tubes.

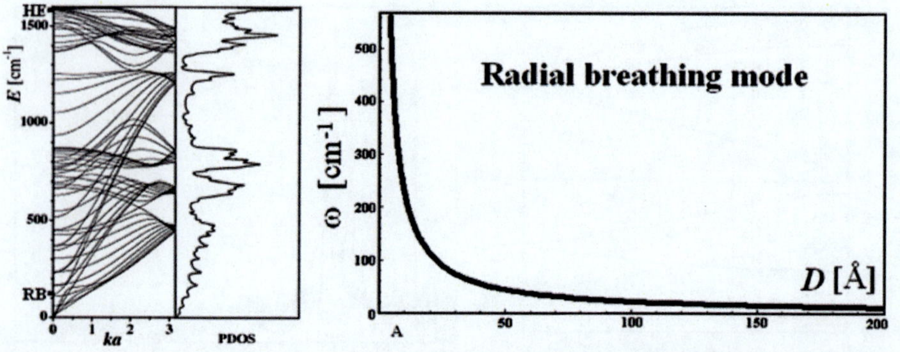

Fig. 3.5 *Left*: Phonon bands of the nanotube (10,10), with radial breathing and high energy modes denoted by RB and HE, and density of states (DOS). *Right*: Dependence of the radial breathing mode frequency of the diameter

It is experimentally found that the totally symmetric modes, corresponding to the representation ${}_0A_0^+$, are much more intensive in the Raman spectra than the others. The lowest among them is called breathing mode, because the vibrations of the atoms are almost radial, and in the limiting case of infinite diameter it becomes transversal acoustic graphene mode (Fig. 3.5). The other two (for chiral) or one (for achiral tubes) symmetric modes are in the high-energy region. The most important is the fact that the frequency of this mode decreases with diameter as $\omega[\mathrm{cm}^{-1}] = 2243/D[\text{Å}]$. Raman measurements [7], being a frequent and cheep experimental tool, can be used to characterize the samples.

3.3 Interaction Between Walls

In this section, the friction between the walls in multi-wall tubes will be considered. It should be emphasized that the amazing conclusion of the almost free-sliding and twisting walls is a direct consequence of symmetry, and therefore qualitatively independent on the underlying potential $v(|\boldsymbol{r}_1 - \boldsymbol{r}_2|)$ of the interaction between the carbon atoms. Also, it should be noted that the role of the interior and exterior nanotube is completely equivalent, and the interior one is considered firstly only for clarity.

3.3.1 Potential Produced by Nanotube

It is natural to assume that the interaction between walls is the sum of the pairwise interatomic potentials $v(|\boldsymbol{r}_1 - \boldsymbol{r}_2|)$. Then, it is possible to find the potential produced at the point $\boldsymbol{r}$ by the interior nanotube as the sum over its atoms C_{tsu}:

$$V_{\text{in}}(\boldsymbol{r}) = \sum_{tsu} v(\boldsymbol{r}, \boldsymbol{r}_{tsu}). \tag{3.6}$$

This resulting potential is invariant under any transformation of symmetry, as it is manifested only as a permutation of terms in sum. Therefore, this potential may be expanded as a series of invariant functions of the symmetry group of the nanotube (n_1, n_2). It is easy to show that a basis of invariant functions is:

$$F^M_{KR}(\rho, \varphi, z) = F_R(\rho) \cos\left(nM\varphi + 2\pi\left(K - \frac{Mnr}{q}\right) z/f\right); \tag{3.7}$$

here $K, R = 0, \pm 1, \ldots, M = 0, 1, \ldots$, and $F_R(\rho)$ is arbitrary basis in the space of functions over ρ (usually Bessel functions are used). Therefore, the mentioned expansion at the fixed radius ρ (e.g., radius of the outer wall) is:

$$V_{\text{in}}(\boldsymbol{r}) = \sum_{KM} \alpha^M_K(\rho) \cos\left(nM\varphi + 2\pi\left(K - \frac{Mnr}{q}\right) z/f\right). \tag{3.8}$$

Note that only the terms with $|K| = M = 1$ have the symmetry of the nanotube, while other ones have larger symmetry (i.e., the other terms are invariant under additional transformations). This means that if these terms vanish, the potential has larger symmetry than the tube. This is not physically allowed, as the experiment would show larger symmetry than it really is. Numerically, independently on the pairwise potential used (usually pairwise interaction is modeled by Lenard-Jones potential), it appears that these terms (i.e., their coefficients α^M_K are by far greater then the others (Fig. 3.6)).

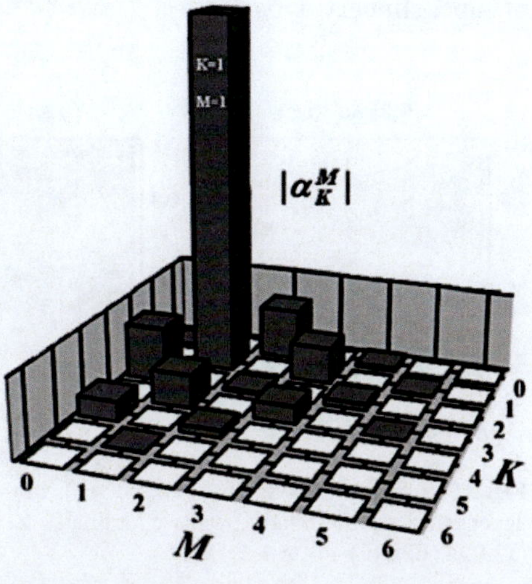

Fig. 3.6 Expansion amplitudes α^M_ω of the potential $V_{\text{in}}(\varphi, z)$ of (12,12) at ρ corresponding to the outer wall, below

3.3.2 Interaction

As it is anticipated, the total interaction potential is the sum over pairwise contributions:

$$V(\Phi, Z) = \sum_{t's'u'} \sum_{tsu} v(\boldsymbol{r}'_{t's'u'}, \boldsymbol{r}_{tsu}) = \sum_{t',s',u'} V_{\text{in}}(\boldsymbol{r}'_{t's'u'}). \tag{3.9}$$

It can be observed that the interaction potential is again sum over atoms of the inner and outer wall, i.e., it is invariant under both groups of symmetry, $\boldsymbol{L}_W$ and $\boldsymbol{L}_{W'}$. Therefore, its symmetry group is a product of the symmetry groups divided by their intersection (to avoid multiple counting of the common elements): $\boldsymbol{L}_{\text{int}} = \boldsymbol{L}\boldsymbol{L}'/\boldsymbol{L} \cap \boldsymbol{L}'$. If the groups were the same ($\boldsymbol{L} = \boldsymbol{L}'$), then the intersection would be the same group, $\boldsymbol{L} \cap \boldsymbol{L}' = \boldsymbol{L}$, as well as the final group $\boldsymbol{L}_{\text{int}} = \boldsymbol{L}$. However, $\boldsymbol{L}$ and $\boldsymbol{L}'$ are highly incompatible, making the intersection very restricted and consequently $\boldsymbol{L}_{\text{int}}$ is very large. After this conclusion, it is clear that the large symmetry of the interacting potential implies that when (3.8) is substituted in (3.9), the sum over outer wall atoms will leave only the terms which are invariant also under the transformations of $\boldsymbol{L}_{W'}$. This property have only terms with large K and M. However, their amplitudes are very small, leading to the negligible contribution to the total interaction. This proves that friction between walls is very small. Numerical results, with Lenard-Jones potential are in Fig. 3.7.

In particular, for incommensurate walls, $\boldsymbol{L} \cap \boldsymbol{L}'$ is finite. Then the group $\boldsymbol{L}_{\text{int}}$ of the symmetry of the potential will map any point into the quasicontinual set of the points with the same potential energy. Therefore, quasicontinually many values of the relative coordinate Z have the same energy. On the other hand, no discontinuity in the energy can occur during summation. Altogether, one concludes that the potential is constant along Z. Consequently, no energy, and thereafter, force is needed to perform sliding of one wall relatively to another, and this is an example of super slippery dynamics.

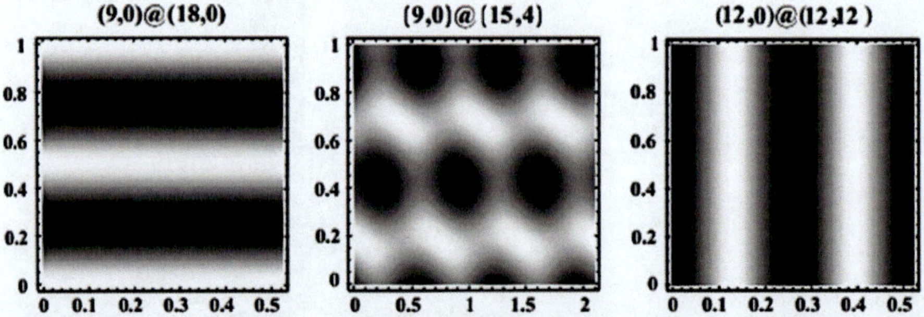

Fig. 3.7 Interaction potential as of the walls in the double-wall nanotubes. Density plot presents dependence on the relative walls coordinates Z and Φ. Note smoothness along Z in the case (12,0)@(12@12)

Of course, all this is valid for ideal, coaxially arranged infinite walls. Real samples are hardly defect free, and always are finite. Therefore, really there is no super slippery cases, but it is explained why experimentally measured friction is extremely low. This effect inspired a number of projects of nanomachines using double-wall tubes as a moving parts [11, 12].

3.4 Summary

Nanotubes are among the most interesting challenges of the contemporary science and technology. One of the most profound ways to understand them, and efficiently predict their properties is symmetry, described in terms of the line groups. In particular, amazing conductivity, optical properties and low friction between walls are direct consequence of symmetry.

References

1. S. Iijima, Nature **354**, 56 (1991)
2. *Applied Physics of Nanotubes: Fundamentals of Theory, Optics and Transport Devices*, ed. by S.V. Rotkin, S. Subramoney (Springer-Verlag, Berlin-Heidelberg-New York, 2005)
3. N. Hamada, S. Sawada, A. Oshiyama, Phys. Rev. Lett. **68**, 1579 (1992)
4. C. T. White, D. H. Robertson, J. W. Mintmire, Phys. Rev. B **47**, 5485 (1993)
5. J. W. Mintmire, C. T. White, Phys. Rev. Lett. **81**, 2506 (1998)
6. M. Damnjanović, I. Milošević, T. Vuković, R. Sredanović, Phys. Rev. B **60**, 2728–2739 (1999)
7. S. Reich, C. Thomsen, J. Maultzsch, *Carbon Nanotubes – Basic Concepts and Physical Properties* (Wiley-VCH, Weinheim, 2004)
8. E.B. Barros, A. Jorio, G.G. Samsonidze, R.B. Capaz, A.G. Souza Filho, J.M. Filho, G. Dresselhaus, Review on the symmetry-related properties of carbon nanotubes. Phys. Rep., **431**, 261–302 (2006)
9. K.S. Novoselov, A.K. Geim, S.V. Morozov, D. Jiang, Y. Zhang, S.V. Dubonos, I.V. Grigorieva, A. A. Firsov, Science **306**, 666 (2004)
10. L.D. Landau, E.M. Lifshitz, *Quantum Mechanics* (Elsevier, Oxford, 1980)
11. E. Bichoutskaia, A. Popov, Y. Lozovik, G. Ivanchenko, N. Lebedev, Phys. Lett. A **366**, 480–486 (2007)
12. X.G. Zhao, P. T. Cummings, J. Chem. Phys. **124**, 134705 (2006)

Chapter 4
Laser-Based Growth of Nanostructured Thin Films

P. Patsalas

Abstract The development of powerful, pulsed lasers with immense power has drastically changed our perception of light-matter interactions and opened new ways of implementing laser sources for the growth and processing of nanostructured materials, making Pulsed Laser Deposition (PLD) as one of the most important techniques in the nanotechnology era. In this work, we describe the main parts of a PLD system and the basic physical processes involved, as well as some laser processes for microstructural control of the grown materials. In order to establish firm understanding of the PLD processes, three case studies are presented as examples: (a) External Control of Ablated Species and Application to Tetrahedral Amorphous Carbon (ta-C) Films, (b) Self-assembled nanoparticles (NPs) into dielectric-matrix films and superlattices, (c) Controlling of the atomic structure and nanostructure of intermetallic and glassy films.

4.1 Introduction

The development of powerful, high photon flux, Q-switched lasers has drastically changed our perception of light-matter interactions and opened new ways of implementing laser sources for the growth and processing of nanostructured materials. Therefore, Pulsed Laser Deposition (PLD) has emerged as a very important growth technique in the nanotechnology era. PLD has become a well-established technique for the growth of carbon nanotubes [1–21], diamond like carbon and ultra-nano-crystalline diamond [22–45], nanocomposite films and coatings with finely controlled dispersion of nanoparticles (NPs) and superlattices [45–68], and many

P. Patsalas (✉)
Department of Materials Science and Engineering, University of Ioannina,
GR-45110 Ioannina, Greece
e-mail: ppats@cc.uoi.gr

S. Logothetidis (ed.), *Nanostructured Materials and Their Applications*,
NanoScience and Technology, DOI 10.1007/978-3-642-22227-6_4,

other forms of nano-materials such as magnetic, superconducting, and high-k oxide films [69–73]. In addition to PLD, other materials' processes that are based on lasers include, but are not limited to, laser-assisted chemical vapor deposition of patterned materials [74–85], laser annealing and pattering of optoelectronic materials and devices [86–103], and production of NPs by ablation in liquid environments [104–116].

The wide use of PLD is based on its unique combination of assets such as:

1. Clean character, since usually no carrier or precursor gas is required for PLD; the cleanliness of PLD is usually comparable to molecular beam epitaxy (MBE).
2. Retaining the targets composition in the grown film, unlike sputtering and evaporation.
3. Fine control of the kinetic energy of the deposited species.
4. Extremely high deposition rate and nucleation density during the laser pulse (although the effective deposition rate is usually very low due to the pulsed mode, which incorporates immense dead times), which as a result alter the kinetic and thermodynamic conditions of growth.

In this review, we discuss the main parts of a PLD system and the basic physical processes involved, as well as some laser processes used for the control of the microstructure and of the properties of materials. Special emphasis will be given to the control of the kinetic energy of the ablated species as well as to the methodologies employed for the production of either nanocomposite metal-ceramic and intermetallic coatings or single-phase glassy films; these processes will be illustrated by three characteristic case studies of PLD growth: (1) tetrahedral amorphous Carbon (ta-C), (2) AlN:Ag nanocomposites, and (3) Zr–Cu intermetallic films.

4.2 Instrumentation and Principles of Pulsed Laser Deposition

The basic instrument for all laser processes of materials is the laser source itself. The laser source consists of a power source used for the optical pumping, which might be electricity or a strong light source (e.g., the flash lamps of the typical Nd:YAG solid state lasers or another laser source), the active medium and the resonator (or cavity). The active medium can be either a mixture of gases (e.g., He–Ne) or liquids (e.g., various dyes) or a solid state crystal with a controlled concentration of optically active impurities, e.g., Nd impurities in an Yttrium–Aluminum Garnet for Nd:YAG lasers or Ti color centers into an Al_2O_3 crystal for Ti:Sapphire lasers. The stimulated optical emission takes place into the active medium. The resonator is usually a tube with two assembled mirrors. In the case of continuous wave (CW) lasers the one mirror is highly reflective and the other is semi-reflective (the latter is also called the aperture) and the active medium is located between them. The stimulated optical emission occurs along the central axis of the resonator. The generated laser light is

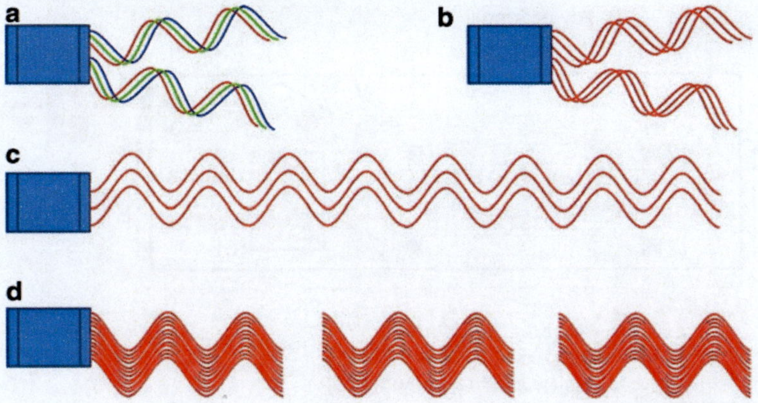

Fig. 4.1 Four types of photon sources: (**a**) a conventional white light source (e.g., fluorescent light, Xe-lamp, etc.), which emits polychromatic and incoherent light, (**b**) a monochromatic source (e.g., a light emitting diode), which emits a single-color but incoherent light, (**c**) a CW-laser, which emits continuously monochromatic, highly directional and coherent light, and (**d**) a Q-switched pulsed laser, which emits pulses of monochromatic, highly directional and coherent light of immense power

monochromatic (single color), coherent (the laser photons share the same phase), and highly directional (almost parallel beam), unlike the conventional light sources, as shown in Fig. 4.1.

The laser sources can be categorized to CW and to pulsed Q-switched lasers. The CW-lasers operate resonators with one partially reflective (~50–80%) mirror out of which a continuous optical beam is emitted continuously, see Fig. 4.1c. On the other hand, a Q-switched laser, Fig. 4.2, is based on the introduction of an electro-optic or acousto-optic modulator (e.g., Kerr cell, Pockel cell, etc.) intersecting the central axis of the resonator.

The modulator for low-Q (low quality factor) conditions is transparent to the emitted laser light along the central axis of the resonator while for high-Q conditions transmits the emitted photons to another direction and through an aperture the laser beam is emitted out of the resonator. In this mode of operation, the modulator is in low-Q conditions, for most of the time, building the appropriate intensity of the laser beam through successive passages of the emitted photons through the active medium. When the desired laser intensity is built, the modulator is switched to high-Q conditions. This results in high peak power (usually in the range 10^8–10^{16} Watt/cm^2) as the average power of the laser is packed into an ultra-short time frame and, thus, a laser pulse of immense power is emitted through the aperture. The pulse duration can range from several tens of ns (10^{-9} s) to a few tens of fs (10^{-15} s). Special crystals can be adapted on the aperture of the laser source in order to generate high harmonics of the light and, thus, emit laser beams of various wavelengths (e.g., for the most popular Nd:YAG lasers the emitted wavelengths can be 1,064, 532, 355, 266, and 213 nm).

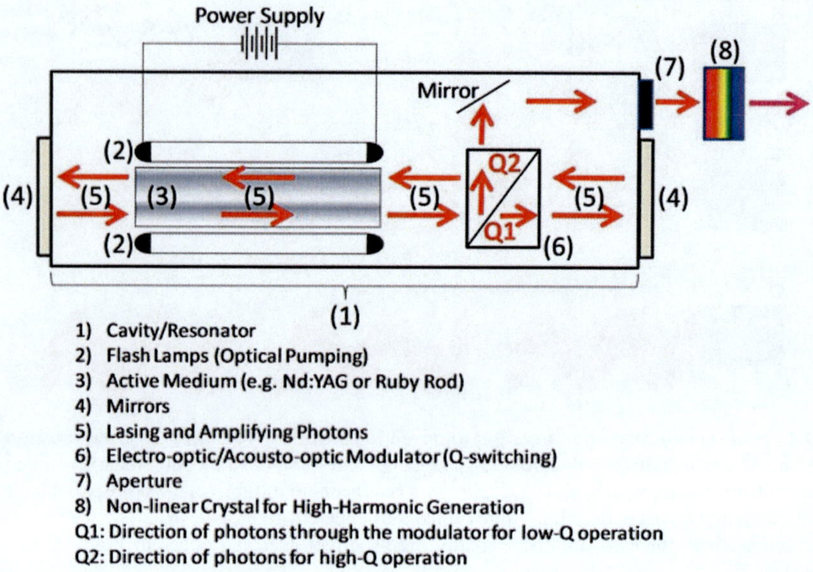

Fig. 4.2 An internal view of a Q-switched pulsed laser source used for PLD. Note that the geometry of the Q-switch modulator depends on the type of modulator (e.g., Kerr cell, Pockel cell, etc.); an oversimplified sketch of the modulator is shown here just for demonstration of its use

Table 4.1 The most common types of lasers used for PLD and their basic features

	Laser type	Pulse duration	Wavelength (nm)	Spectral region	Comments
1	Excimer ArF	Ns	193	FUV	High power Relatively large area
2	Excimer KrF	Ns	248	UV	High power Relatively large area
3	Nd:YAG	ns-ps	1,064 (fundamental) 532 (2nd harmonic) 355 (3rd harmonic) 266 (4th harmonic) 213 (5th harmonic)	IR Green UV UV UV	Compact Robust Stable Many wavelengths Cost effective User friendly
4	Ti:Sapphire	Fs	800 (fundamental) 400 (2nd harmonic)	NIR Violet	Ultra-short pulses Immense power Very expensive

PLD employs exclusively Q-switched pulsed lasers, since only such lasers can provide the required power for the ablation process. The most popular Q-switched pulsed lasers used for PLD and their basic features are displayed in Table 4.1.

PLD is one of the most controllable and versatile lab-scale growth techniques available, although it usually grows samples of small area and only on flat substrates.

PLD is based on the phenomenon of laser ablation in which a focused pulsed laser beam of high fluence illuminates the target material; the ablation process has been reviewed in detail in [117]. The usual fluence used in PLD with pulse duration of the order of ns ranges from a few mJ/cm^2 to several tens of J/cm^2. The fluence when shorter laser pulses (e.g., ps or fs range) are used is lower but the radiation power is still higher. The used fluence is usually equivalent to 1–100 billion times the power density of arriving sunlight on the surface of the earth for the pulse duration and for the specific illumination area. The laser irradiation induces vaporization, via heating of the target, and formation of plasma via ionization of the target atoms. In particular, the removal of atoms from the bulk material is usually done by a Coulomb explosion due to multi-photon ionization of near-surface atoms, given that the laser fluence is of the order of some J/cm^2 (high fluence is a prerequisite for this process, since it ensures a high probability of multi-photon ionization, which is taking place in a time interval of the order of few ps). Subsequently the electrons that become free after the ionization process oscillate within the electromagnetic field of the laser (given that the pulse duration is relatively longer than the ionization process, i.e., pulse duration in the range of ns) and can interact with the target atoms inducing electron–phonon interactions resulting to target heating and vaporization. It is, then, well understood that if shorter wavelength is used (i.e., fs laser pulses) these thermal phenomena will be avoided.

The mixed vapors and ions of the target material are called the plume. The kinetic energy of the ablated species may vary with the laser wavelength and fluence in the range from a few eVs up to hundreds of eVs [29]. After the creation of the plume the material expands within a cone, whose axis is parallel to the normal vector of the target surface toward the substrate due to Coulomb repulsion (for ions) and adiabatic expansion of the pressurized vapors. These processes are displayed schematically in Fig. 4.3.

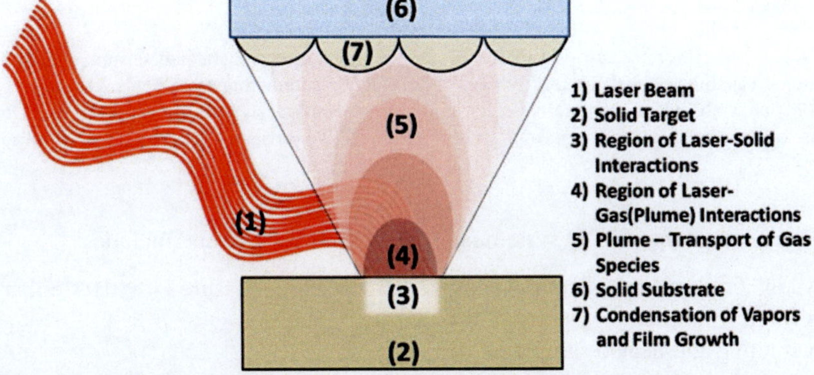

Fig. 4.3 A schematic of the laser ablation process and its stages up to thin film formation

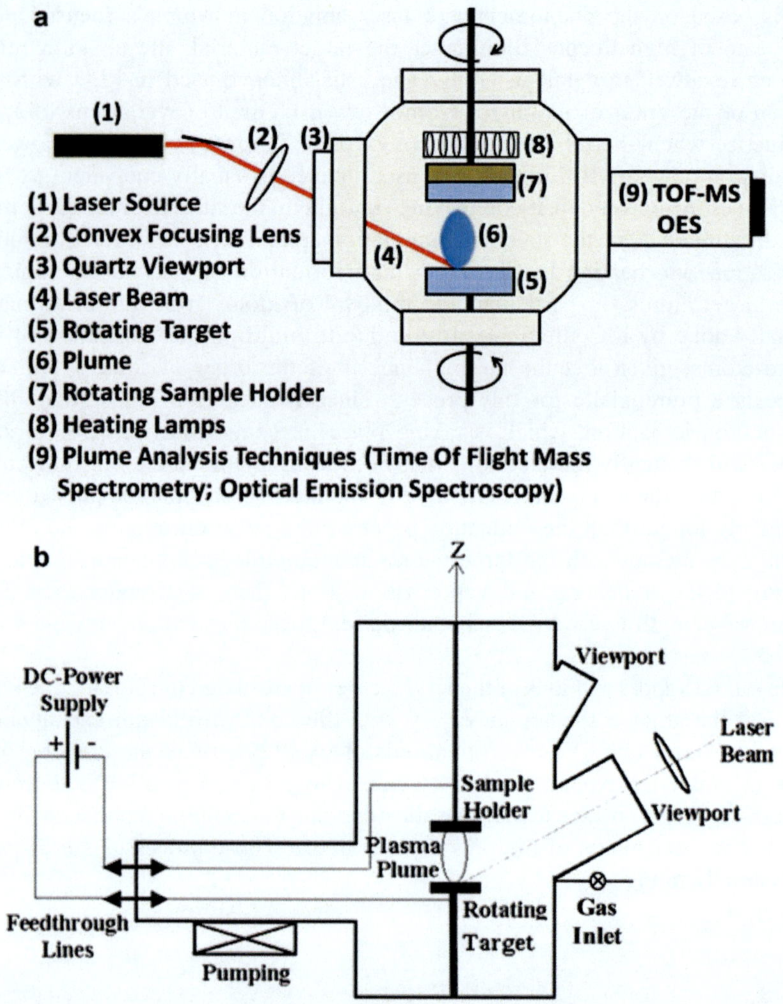

Fig. 4.4 (**a**) The basic set-up of a PLD system. Various more sophisticated set-ups, which offer additional capabilities such as reactive gas flow, electric and/or magnetic fields, etc., also exist like the one of (**b**) where the external electric field is specially designed for controlling the kinetic energy of ionic species and for plasma generation in reactive processes for the growth of nitrides

The basic set up of PLD system is presented in Fig. 4.4a and includes:

- A high or ultra-high vacuum chamber (usually base pressure $P_b < 10^{-7}$ mbar)
- A rotating target
- A rotating and heated sample holder
- A quartz viewport for introducing the laser
- A Q-switched pulsed laser source

Other more sophisticated set-ups, which offer additional capabilities such as reactive gas flow, electric and/or magnetic field also exist like the one presented in Fig. 4.4b. In the latter case [29], the external electric field is specially designed for controlling the kinetic energy of ionic species and for plasma generation in reactive processes for the growth of nitrides.

PLD combines many assets of thermal techniques (such as MBE and evaporation) like the clean character, and of ionic techniques (such as sputtering) like the flexibility in controlling the kinetic energy of the deposited species. The main advantages of PLD can be summarized as:

Table 4.2 Comparison of PLD with the most widely used thin film growth techniques

Technique	Disadvantage	Comment/example	PLD
Thermal techniques (MBE, MOCVD, LPCVD)	Hard to grow metastable phases	No DLC growth	Can grow DLC and other metastable phases
	Hard to handle and control reactive O_2	Hard to grow stoichiometric oxides, perovskites, ferrites	Can grow these materials using oxide target
	Substrate limitations	No RT growth Temperature-sensitive substrates, like organics, cannot be used	Can be used for any substrate
Sputtering	Preferential sputtering of complex targets	Does not retain target composition (in a single-magnetron configuration)	PLD is the best technique to retain the target composition into the film
	Target poisoning	Difficult to control reactive processes	Laser ablates all forms of targets
Vacuum arc deposition (VAD)	Target limitations	Only conducting targets; no oxides, no organics	Laser ablates all forms of targets
	Side effects	Droplets, Thickness Inhomogeneity	Similar
Sputtering/PECVD	Plasma impurities	Predominantly Ar	No impurities, since no carrier gas is used
Sputtering/PECVD/VAD	Internal stress	Higher in subplantation mode	Less than that of the competition
PECVD	Precursor impurities	Organic or halide residues; higher resistivity of nitride films compared to PVD	No impurities

- Flexibility in creating high/low-energy species,
- Low working pressures resulting to higher diffusion of arriving species and subsequently to better crystalline quality of the produced films,
- Ion–solid interactions (implantation, cascades, etc.),
- Clean character; no impurities,
- Growth of metastable phases.

However, PLD has also some drawbacks like:

- Thickness inhomogeneity (i.e., smaller samples compared to other physical or chemical vapor deposition techniques),
- Relatively high surface roughness (droplets/clusters),
- Low-effective deposition rate (few nm/min).

A short comparison of PLD with other thin film growth techniques is presented in Table 4.2.

We should point out that although the apparent deposition rate for PLD is very low (compared to other Physical Vapor Deposition –PVD-techniques, such as sputtering) and it is usually of the order of few nm/min, the real deposition rate is in the order of magnitudes higher because of the pulsed character of deposition (see Fig. 4.5). Thus, the deposition takes place in a time interval which is comparable to the pulse duration (a few ns, or ps, or fs), followed by a dead time which is in the range of 0.1–100 ms for 10 Hz to 10 kHz repetition rate. In conclusion the real deposition rate during the laser pulse is about 10^{-2} nm/pulse, which is equivalent to about 10^6 nm/s and it is much higher than any other PVD technique. The real deposition rate is not so significant for the industrial applications (where the apparent or average deposition rate is the key parameter). On the other hand, the immense real deposition rate is very important for the kinetics and thermodynamics of growth, especially for multi-component films.

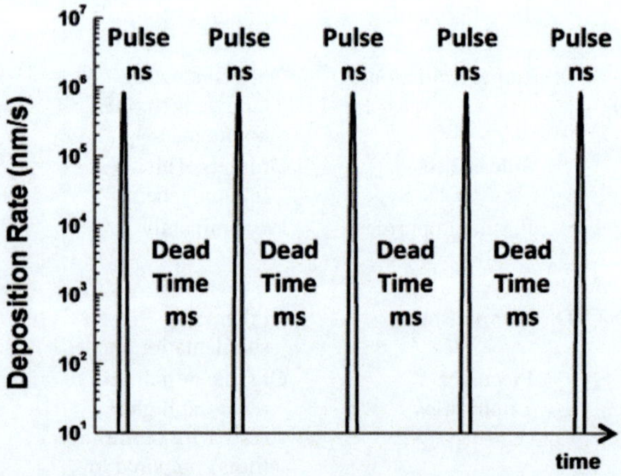

Fig. 4.5 A schematic of the pulsed mode of deposition, which results to a succession of immense deposition rate during the pulse followed by a dead time

4.3 Examples and Applications

4.3.1 External Control of Ablated Species and Application to Ta-C Films [29]

PLD using short wavelengths (UV, FUV range) is one of the most successful techniques for the growth of carbon-based films [22–45, 118, 119], among them tetrahedral a-C (ta-C). Here we implement another approach in PLD, based on the use of a static electric field without any other change in the working conditions. We present the PLD of ta-C as a case study of this approach. ta-C is a very important engineering material finding several industrial applications [120]. Conventional PLD of ta-C is exclusively performed using short laser wavelengths generated by ArF (=193 nm) or KrF (=248 nm) excimer lasers or the high-order (>3rd) harmonics of Nd:YAG contributing to the drawbacks for industrial scale-up. The short laser wavelengths ($\lambda < 400$ nm) are more efficient for the growth of ta-C mainly because of the higher ion/neutral ratio produced in the plume. In most of the PLD experiments the growth of ta-C is achieved by manipulating the plume characteristics through variation of the laser fluence. It is widely accepted that high fluence (some tens of $\mathrm{J/cm^2}$) are required to grow high-quality ta-C [39,42,45,51]. However, at high fluence processes there is a competitive destructive mechanism from the generation of the heavy ablated species/clusters [121, 122] that are finally degrading the ta-C quality (i.e., sp^3 content and surface roughening). The reported application of a dc bias to the substrate during PLD growth of ta-C did not result to a substantial increase but rather to a decrease, when short wavelength was used [123]. We have confirmed this [29] for the UV ablation of C, however, we show here that the use of an external static electric field enables the growth of high quality ta-C films using just the second harmonic ($\lambda = 532$ nm, Green) of a Nd:YAG laser, a fact with implications in the industrial implementation of PLD for ta-C growth.

The first (1,064 nm), second (532 nm), or third (355 nm) harmonics of a Nd:YAG laser source, pulse duration of 3 ns, repetition rate of 10 Hz, were used to ablate the graphite target in high vacuum (base pressure $P_b = 5 \times 10^{-5}$ Pa). The beam was focused outside the vacuum chamber passing through a fused silica viewport via a lens (50 cm focal length). In all experiments the laser fluence was kept constant at 24, 60, and 90 $\mathrm{J/cm^2}$ for the first, second, and third harmonics, respectively [29]. The insulated target and sample holder were electrically connected to a DC power supply. Thus the sample holder is in negative potential relatively to the graphite target. This electrical circuit prevents the electron radiation of the sample and orients and accelerates the various ion species (C^+, C^{2+}, dimmers $C_2{}^+$, trimers $C_3{}^+$, etc.) toward the sample surface.

The use of the static electric field is expected to be beneficial to PLD growth as (1) it prevents the electron radiation of the sample during growth (something that cannot be achieved by applying RF electric field), and (2) it orients and accelerates to the desired kinetic energy the carbon ions toward the substrate. In addition, this

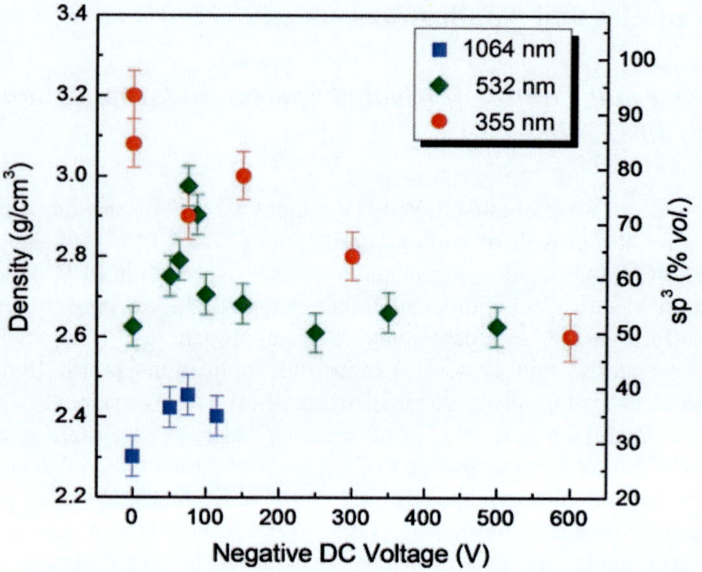

Fig. 4.6 The variation of the density and the sp^3 content vs. the applied external electric field for PLD using the fundamental line and the second and third harmonics of the Nd:YAG laser

experimental setup can provide a more flexible variation of the kinetic energy of the depositing species, rather independently from the laser fluence.

The density of the ta-C films grown without the application of an external electric field is typical of ta-C [45,120]. The denser film has been produced by ablation using the $\lambda = 355$ nm and exhibits density $3.22 \pm 0.05\,\mathrm{g/cm^3}$. The variation of film density, determined by X-Ray Reflectivity (XRR) [29] vs. the applied static electric field are presented in Fig. 4.6 for $\lambda = 1,064$, 532, and 355 nm. The sp^3 content in the films can be also determined using the density-sp^3 correlation [124]. There is a strikingly different variation between films grown using different laser wavelengths. All the films grown using $\lambda = 1,064$ nm are mostly graphitic (sp^3 $< 50\%$), while the films grown with $\lambda = 355$ nm are all predominantly tetrahedral (sp^3 $> 50\%$). The variations with the external electric field are more pronounced in the case of $\lambda = 532$ nm. The density variations can be well understood taking into account the ionization conditions of carbon and the ablated species produced in each case (even without the application of the external field). The ionization energy of atomic carbon is 11.25 eV.

This means that the absorption of four photons, from the same laser pulse, at 355 nm, leads to carbon ionization, while for carbon ionization at 532 and 1,064 nm five- and ten-photon absorption processes, respectively, should be involved. This implies that, for the same laser intensity, the ionization efficiency is much higher at 355 nm. As a consequence, the relative abundance of the ionic species in the plume is expected to be much higher at 355 nm than that at 532 nm. Therefore, the Coulomb repulsion between the ionic species in the plume at 355 nm results in ions with significant kinetic energies, while the actual value depends on the experimental parameters of the ablation process. At $\lambda = 355$ nm and for the higher fluence used in our experiments the dominant species are expected to be single-charged carbon ions, C^+ [125]. The present experimental data indicate that the kinetic energy of these C^+ ions at 355 nm (without the applied electric field) should be tens of eV's, in order to form such dense ta-C [120]. On the contrary, deposition of species with thermal kinetic energies is gradually deteriorating the film's density, as observed in the case of $\lambda = 532$ and 1,064 nm (Fig. 4.6). In this case, the ablation of graphite using high fluence of photons with $\lambda = 532$ nm produces heavy C clusters, which may be formed also in the gas phase [122]. The production of such heavy and slow species is one of the most important problems of PLD growth of ta-C because they contribute to sp^2 bonding. Finally, the ablated carbon atoms are very improbable to be in ionic form in the case of $\lambda = 1,064$ nm, thus explaining the low density and graphitic character of these films.

The variation of the density of the produced films with the wavelength is clearly illustrated in Fig. 4.7 and can be well understood after the previous discussion. The effect of the laser fluence is presented by comparing [29] with the results presented by Yamamoto et al. [126] for a fixed fluence of $2\,\mathrm{J/cm^2}$.

When an external electric field is applied, the ablation process remains unaffected but the plume composition may change drastically, given that there is sufficient concentration of ionic species. Otherwise, the effect of the external field is minor, as in the case of $\lambda = 1,064$ nm. Indeed, the experimental data using the 1,064 nm wavelength indicate that the increase of density (which is associated to the fraction of ions in the plume) is very weak, namely, from 2.30 to $2.45\,\mathrm{g/cm^3}$. This is determined to be equivalent to a small increase (10.5%) of sp^3 content, using the formula of Ferrari et al. [124]. The influence on the plume synthesis is more

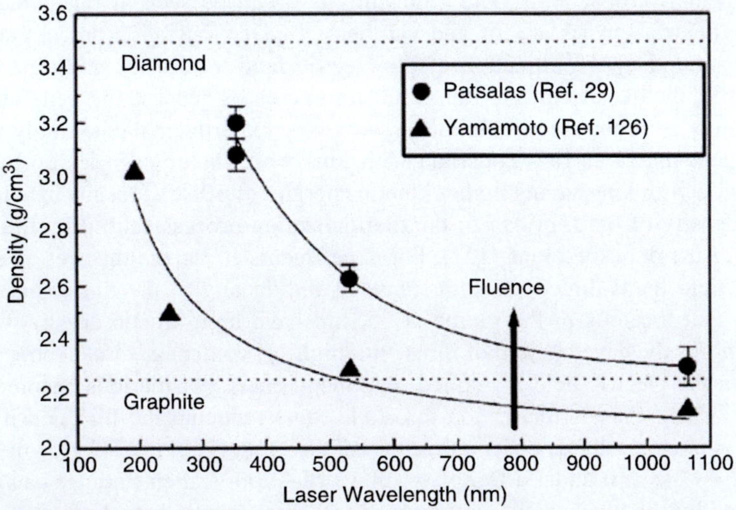

Fig. 4.7 The density of the ta-C films grown by PLD vs. the wavelength used for the ablation

conspicuous at 532 nm. In this case, the small abundance of the ions (compared to 355 nm), in the absence of the field, increases due to the ionization of the neutral species via impact with the ions accelerated by the field, following gas-phase reactions, which incorporate fast C^+ [29]. It is worth commenting that the maximum density values of the films grown using 1,064 and 532 nm lasers are observed at the same external electric field ($V_b = -75$ V), although the ablated species for the 1,064 and 532 nm lasers have different mean initial kinetic energies E_k, as indicated by the density values of the films without the application of the electric field (Fig. 4.6). This clearly illustrates that the major effect of the external field to the growth mechanism is the promotion of the secondary ionization of the ablated species through collisions in the gas phase [29] resulting to higher ion density.

The maximum density increase occurs at an electric field about 22 eV/cm (75 V bias and 3.5 cm substrate to target distance). This means that an ablated C^+ ion with zero Ek needs to travel about 0.5 cm to gain kinetic energy of 11.25 eV, which is the minimum energy required for the ionization of a C neutral through collision with C^+. However, at this distance the local pressure in the plume is expected to be so low and the mean free path of the gas species so long that the probability of collision and secondary ionization probability would be very small making this mechanism less efficient. This is expected for processes involving species of very small E_k such as the 1,064 nm PLD. When E_k is higher the required distance for a C^+ to gain a total kinetic energy of 11.25 eV is shorter than 0.5 cm, where the local pressure in the plume is much higher making the secondary ionization more probable and efficient. The mechanism of creating species with high E_k is through Coulomb repulsion and requires the existence of a significant concentration of ions as in the case of 532 nm. Finally, for the 355 nm the abundance of the ionic species is very high at zero field and thus the secondary ionization is very limited.

Evidence on the secondary ionization mechanism is the accretion of the growth rate, especially for $\lambda = 532$ and 1,064 nm, as it can be seen in Fig. 4.8. This is because neutral species are ionized via the secondary ionization process and they are accelerated by the attraction of the electric field toward the substrate. At the same time, the density of the film at 532 nm increases because the generated ions gain energy from the electric field (Fig. 4.6). The experimental data imply that the sp^3 content increases for deposition with ions with kinetic energies up to 80 eV. Deposition with ions having higher kinetic energies (>100 eV) results in a decrease of the density of the film due to the thermalization process caused by the excess energy of the deposited ions [122]. For experiments at 355 nm the presence of the electric field leads directly to a decrease of the density of the film. This finding implies that the ions in the plume at 355 nm have high kinetic energy which is sufficient for the development of films with high sp^3 content at zero electric field.

When the electric field is applied the kinetic energy of the ions becomes even higher passing the thermalization threshold, thus reducing the film density. This finding is in line with the observations of Pappas et al. [122] in ta-C growth by KrF PLD ($\lambda = 248$ nm) under a DC bias voltage of -500 V; their findings can be also explained by the mechanism that we propose. Similar data have been reported by Yamamoto et al. for short wavelengths as well [127].

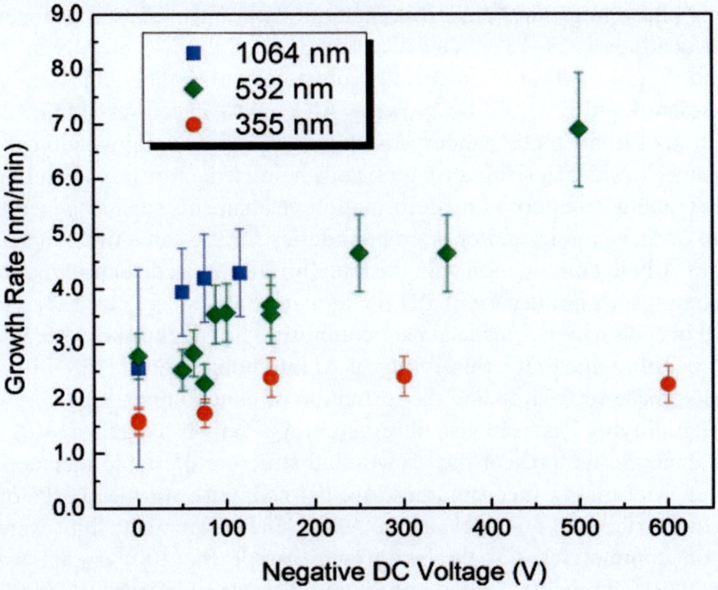

Fig. 4.8 The growth rate vs the applied voltage for the three sets of ta-C films

In this example we have demonstrated the potential of PLD growth combined with the application of homogeneous static (DC) electric field between the PLD target and the substrate for ta-C films. The effects of the electric field are dependent on the laser wavelength used for the ablation and, thus, to the kinds of the ablated species in each case. This is attributed to secondary ionization through gas phase collisions occurring in the plume that are further promoted by the ions, which are accelerated by the electric field. In the case of the first and third harmonics of the Nd:YAG laser the effects of the electric field are weaker due to the already very low or very high degree of ionization of the ablated species, respectively, making the secondary ionization improbable. On the contrary, for the intermediate case of the second harmonic of Nd:YAG laser a considerable improvement of the PLD process was found. This process was proven to improve ta-C films in terms of density (from 2.60 to 2.95 g/cm^3) and deposition rate (from about 2 to 7 nm/min), especially when the second harmonic was used for the ablation.

4.3.2 *Self-Assembled Nanoparticles into Dielectric-Matrix Films and Superlattices [52, 54]*

Aluminum nitride (AlN) is one of the most well-known very wide band gap compound semiconductors, which has also additional assets such as high hardness, high thermal conductivity, and refractory character [128]. It has been also studied as an alloying phase in group IVb–VIb transition metal nitride-based nanocomposite superhard coatings [129–133]. The disadvantages of the use of AlN are its brittleness and its poor adhesion on various substrates [134]. The adhesion issue has been considered and resolved by growing AlN on Al interlayers [134, 135]. The incorporation of noble metal nanocrystals into AlN can be employed to enhance its plasticity since besides intrinsic structural and chemical factors (e.g., bond strength), a strong effect is also imposed by deformation mechanisms such as generation and movement of dislocations and/or grain boundaries. At the same time, the dielectric character of AlN in combination with the metallic NPs may add extra functionalities to these coatings, as in the case of BN:metal nanocomposites [136, 137]. However, the growth of such AlN-noble metal nanocomposites has not been reported yet in the literature, possibly due to the miscibility of Al into noble metals [138, 139] making the Ag-AlN phase separation and the formation of nanocomposites a very difficult task; similar alloying has been also observed in Ag–Ga—N systems [140].

In this example we present the growth and structure of stable nanocomposites based on an AlN matrix incorporating well-defined, pure Ag metal NPs of a very narrow size distribution. The AlN and the AlN:Ag nanocomposite films were grown by PLD on commercial, Czochralski-grown, *n*-type Si (100) crystal wafers of resistivity 1–10 Ωcm using a rotating sectored disk target of pure (99.999%) solid Al and Ag in a flowing N_2 ambient, as shown in Fig. 4.9. A solid disk of the matrix material (in our example is Al) is the basis of the target. Thin sheets of the second material (to be used as inclusions) are assembled on top of the disk dividing the target into sectors of the two constituent materials. The rotating, sectored disk target concept is based upon the successive and periodic ablation of two individual materials using a single laser, which impinges off the central axis of the target as shown in Fig. 4.9.

The series of AlN:Ag nanocomposites studied in this work is listed in Table 4.3. XRD revealed that the Ag NPs exhibited the *fcc* crystal structure with lattice parameter almost identical to that of pure, unstressed Ag [52, 54]; no diffraction from crystalline AlN has been observed in XRD. The size and the filling ratio of the Ag NPs inside the AlN matrix were controlled by varying the geometry (% Ag) and the target rotation frequency (TRF) of the sectored target. In the High-Resolution

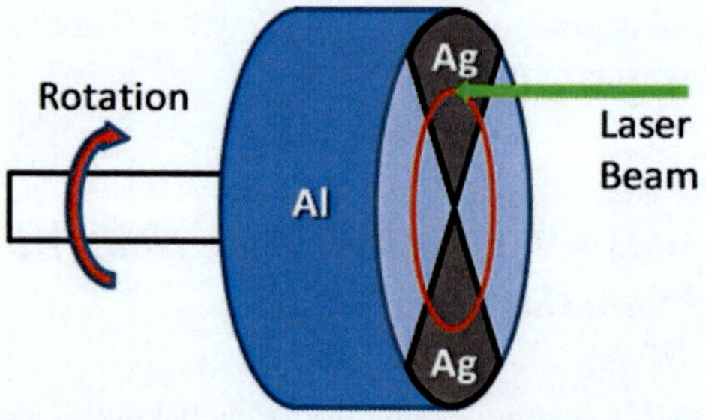

Fig. 4.9 The geometry of the rotating sectored target used for the growth of AlN:Ag nanocomposite films

Table 4.3 List of specimens used in this example

Specimen code	Target composition	Target rotation frequency (Hz)	Pulse energy (mJ)	Working pressure (mbar)	Deposition time (min)
A	Al	0.35	35	3.9×10^{-2}	30
B	Al 87.5%–Ag 12.5%	0.175	35	3.9×10^{-2}	30
C	Al 87.5%–Ag 12.5%	0.0875	35	3.9×10^{-2}	30
D	Al 75%–Ag 25% (Ag2 × 45°)	0.035	35	3.9×10^{-2}	180
E	Al 75%–Ag 25% (Ag4 × 22.5°)	0.035	35	3.9×10^{-2}	40
F	Al 75%–Ag 25% (Ag4 × 22.5°)	0.035	18/35	3.9×10^{-2}	2/35

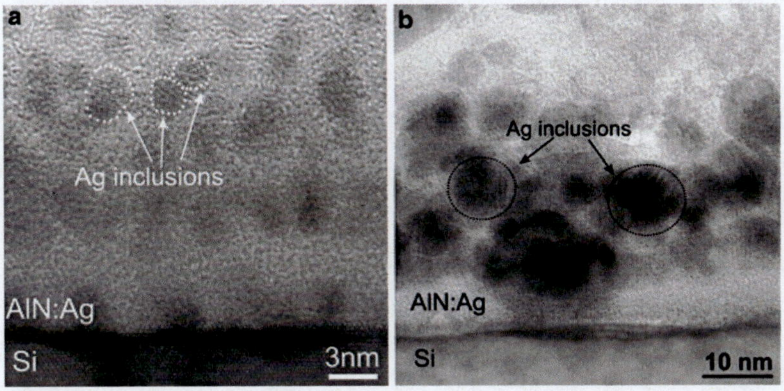

Fig. 4.10 Cross-section HRTEM images from two AlN:Ag samples grown using different width of Ag sectors on the target (**a**) 22.5°, and (**b**) 45°. [Images from [52]]

Transmission Electron Microscopy (HRTEM) images of Fig. 4.10 it can be seen that doubling the angular width of the Ag sectors from 22.5° to 45° results in a significant increase in the mean size of the Ag NPs.

Various target geometries were employed i.e., 12.5% and 25% Ag (sectors of 2 × 22.5°, 2 × 45°, or 4 × 22.5°, as summarized in Table 4.3). The TRF was varied in the range 0.35–0.035 Hz, while the sample holder frequency was constant 0.35 Hz for all samples. The 3rd ($\lambda = 355$ nm) harmonic of a Nd:YAG laser source (pulse duration 3 ns, repetition rate 10 Hz) was used to ablate the target at room temperature (RT). The working pressure of the ablated Al vapors and flowing N_2 under these conditions varied between 1 and 80×10^{-3} mbar. The periodic nature of the ablation process when a rotating sectored target is used enables the growth of nanocomposite films with homogeneous distribution of NPs, as it is observed in Fig. 4.11a. Following this process, the size distribution of the Ag NPs is also extremely narrow according to the size histogram of Fig. 4.11b.

The formation of metal NPs into ceramic matrices by PLD growth is not a special feature of the AlN-Ag system. Similar microstructures have been also observed in a-C:Ag [54, 141], Al_2O_3:Cu [142], and $BaTiO_3$:Au [143] films, as shown in the conventional transmission electron microscopy (TEM) images of Fig. 4.12, showing that this might be a general behavior.

However, Wang et al. reported the growth of TiN/TaN superlattices with very well-defined interfaces (Fig. 4.13) instead of nanocomposites, using similar PLD

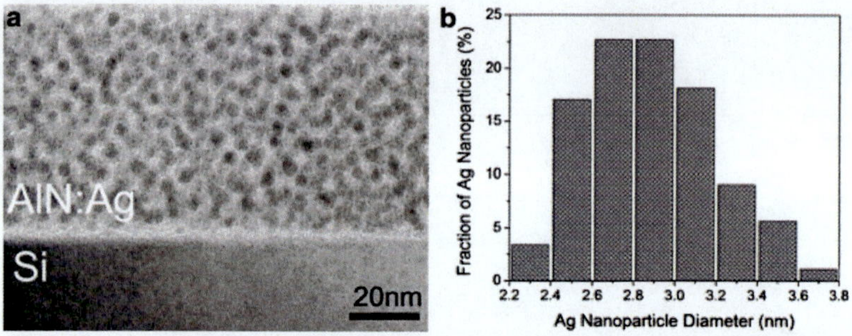

Fig. 4.11 (**a**) Cross-section HRTEM image in geometry from an AlN:Ag ([Ag] = 14.2% at.) nanocomposite film, (**b**) the particle size histogram of the same specimen

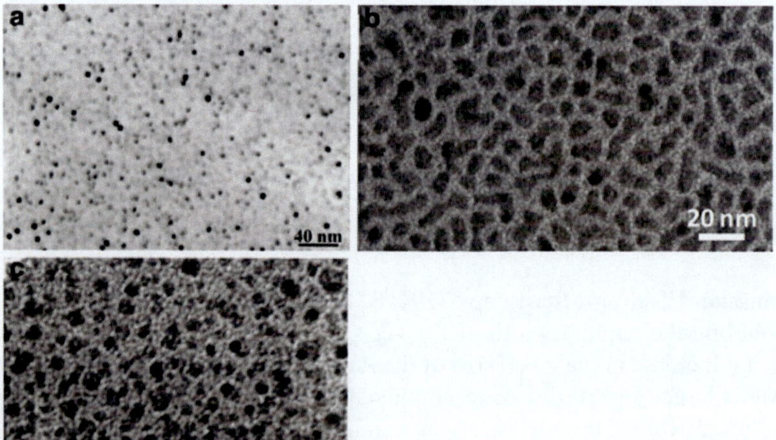

Fig. 4.12 Plan-view TEM images from thin film nanocomposites grown by PLD: (**a**) a-C:Ag [from [54], (**b**) Al_2O_3:Cu [from [142], (**c**) $BaTiO_3$:Au [from [143]. All cases exhibit noble metal NPs embedded into ceramic matrices

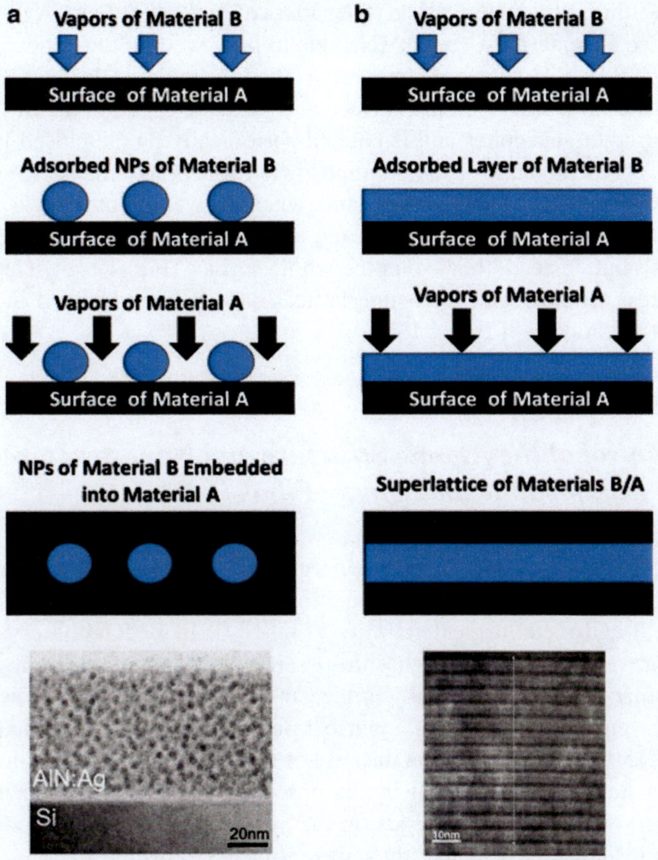

Fig. 4.13 A sketch of the evolution of the growth of (**a**) a nanocomposite AlN:Ag (TEM image from [52,54]), and (**b**) a TiN/TaN (TEM image from [144]), during successive ablation of the two individual materials, which consist the rotating sectored target

conditions, i.e., fluence, pulse duration, repetition rate, rotating sectored target, etc. [144]. This striking difference in the microstructure of the AlN:Ag and TiN/TaN systems, grown by PLD using rotating sectored targets of similar geometry, indicates that the miscibility and wetting of the two constituent phases (either AlN and Ag or TiN and TaN, respectively) plays an important role. TiN and TaN share a similar rocksalt crystal structure, they have a low lattice mismatch (2.7%) and they are completely miscible to each other due to their electronic compatibility [145]. On the other hand, noble metals such as Ag are structurally and electronically incompatible to AlN [146]. Therefore, it is expected that the wetting of TiN on TaN and of Ag on AlN and *vice-versa* would be very good and poor, respectively.

The sketches of Fig. 4.13 depict the different evolution of the growth of the two-phase material during the successive ablation and deposition of the two constituent phases for the cases of poor and good wetting between them, respectively.

When the wetting is poor (Fig. 4.13a) the vapors of material B (e.g., Ag) condensate on the surface of material A (e.g., AlN); due to poor wetting the condensed vapors are forming NPs whose shapes depend on the wetting angle; at the next laser pulses, vapors of material B arrive on the surface and bury the formed NPs of material B. As the sectored target rotates and is ablated periodically by the pulsed laser beam, this process is repeated until a nanocomposite is formed, like the AlN:Ag shown at the bottom of Fig. 4.13a. On the other hand, when the wetting of the two constituent phases is good (like for TiN and TaN) the condensed vapors of the one phase do not form NPs but, instead, they cover the whole surface (Fig. 4.13b). The repetition of this process would result to a superlattice like the one observed in [144] and presented at the bottom of Fig. 4.13b.

4.3.3 Control of the Atomic Structure and Nanostructure of Intermetallic and Glassy Films [147]

Bulk metallic glasses (BMG) have emerged as a very important category of engineering materials due to their combination of exceptional mechanical properties and chemical and metallurgical stability [148–154]. In BMGs the crystallization is usually prevented by using high entropy alloying of many elements; however, glasses of binary systems, especially in the Cu–Zr systems, have been also reported and they are currently subject of intense research due to the simplicity in their production [150, 155]. Focusing on the case of the Cu–Zr archetypical binary BMG system, it is now accepted that it is composed of sub-nm bimetallic icosahedral (ICO) clusters whose nature depends on the system's composition [154–156] and having no translational symmetry, thus preventing crystallization.

For several emerging applications, like the micro-electromechanical systems (MEMS), the metallic glasses should be in the form of thin films, which are usually grown from the vapor phase, thus forming thin film metallic glasses (TFMG). Another advantage of using vapor phase growth for TFMG is its compatibility with the patterning processes (lithography, etching, mask's lift off, etc.) used for MEMS fabrication. The vapor phase growth is a process far away from thermodynamic equilibrium, in which the kinetic effects might be very important in addition to the thermodynamic processes that determine the glass forming ability in BMGs.

PLD has been used as a model film growth technique due to its well-known ability to produce films with homogeneous chemical composition of a multi-elemental target like Cu–Zr. In order to identify the possible effect of the target's structure to the structure of the produced Zr–Cu films we used three types of targets:

1. A $Zr_{70}Cu_{30}$ BMG sheet, produced by the melt-spinning method
2. A homogeneous, polycrystalline intermetallic ($t - Zr_2Cu$) target
3. A sectored target consisting of plates of pure Zr and Cu, with a geometry similar to that shown in Fig. 4.9

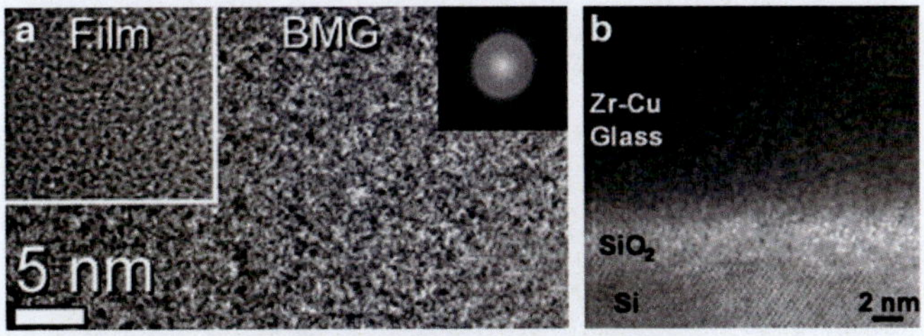

Fig. 4.14 (**a**) Plan-view HRTEM image showing the amorphous structure of the BMG target. The pure glassy state of the Zr–Cu films grown from the BMG ribbon is depicted in plan-view HRTEM image and the corresponding SAED are shown in the insets; the same scale bar applies to both images. (**b**) Cross-section HRTEM image from a Zr–Cu glass film grown from a $t - Zr_2Cu$ target

The plan-view HRTEM images of Fig. 4.14a illustrates the amorphous structure of the BMG target. The Zr–Cu films grown from the $Zr_{70}Cu_{30}$ BMG ribbon are purely glassy and resemble the BMG target, as revealed in the plane view HRTEM image shown as inset in Fig. 4.14a. The structure of the thin film is identical to the BMG target. The pure glassy state of the film has been also confirmed from selected area electron diffraction (SAED), as shown in the upper right side inset of Fig. 4.14a.

A perfect glassy structure of the films grown by PLD from a $t - Zr_2Cu$ polycrystalline target has been also confirmed as shown in the cross-section HRTEM images of Fig. 4.14b. The structural difference between the glassy film and the crystalline Si is evident. The comparison of the films grown by the two different targets demonstrates that the structure of the target does not influence the structure of the films grown by the PLD technique as long as the Zr and Cu species are ablated simultaneously. In both cases the composition of the target, measured by X-Ray Fluorescence, has been retained in the produced films ($Zr_{70}Cu_{30}$ and $t - Zr_2Cu$, respectively), which are in the usual composition range for glassy Zr–Cu [147, 156].

In order to investigate the effects of the adatoms' mobility and surface diffusion we also performed PLD experiments using the rotating, sectored Zr–Cu target. In that case the grown films were not glassy; instead, they exhibited a nanocrystalline structure as shown in the HRTEM plan-view image of Fig. 4.15. The film consists of nanograins of the stable $t - Zr_2Cu$, Fig. 4.15 grain A, and hexagonal $\alpha - Zr$, Fig. 4.15 grain B embedded in an amorphous matrix. No trace of Cu grains has been detected all over the studied area. This confirms experimentally reported molecular dynamics (MD) simulations, which show that Cu diffuses and it is consumed to form Zr_2Cu [147]. According to the MD results in the case of Cu deposition on a ZrCu glass surface the resulting adlayer is mixed exhibiting partial layering and structuring that occurs at the expense of Zr atoms in the BMG. When Zr atoms are deposited on the same BMG surface the mixing is limited close to the interface area, while pure Zr adlayers crystallize in the energetically favored (111) face. Amazingly this effect occurs well below the glass transition temperature and it can be observed even at RT, clearly demonstrating that the surface diffusion of adatoms is significant even at RT. The corresponding experimental findings (Fig. 4.15) are in very good agreement with the MD results.

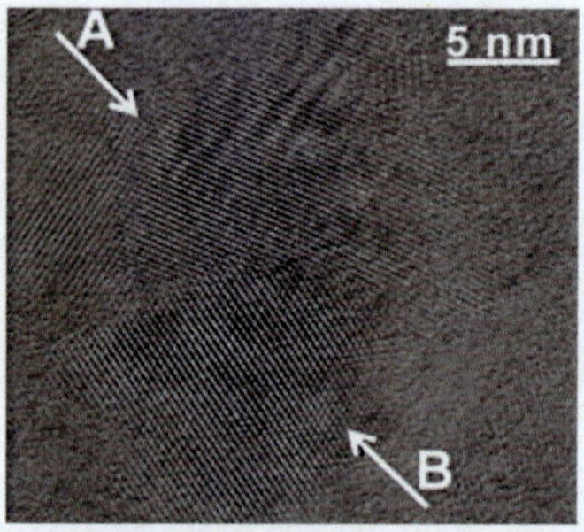

Fig. 4.15 The nanocrystalline structure of a Zr–Cu film grown using a target consisting of Zr and Cu plates (sequential deposition); region A: a tetragonal C11b Zr_2Cu (101) grain; region B: a hexagonal $\alpha - \mathrm{Zr}$ (100) nanograin

We should also point out that the self-organization of the Zr and Cu adatoms observed in the experiment takes place for immensely high real deposition rate. Although the apparent deposition rate for PLD is very low (compared to other PVD techniques, such as sputtering) and in our case is of the order of 6 nm/min, the real deposition rate is in the order of magnitudes higher because of the pulsed character of deposition. Thus, the deposition takes place in a time interval which is comparable to the pulse duration (4 ns), followed by a dead time which is 100 ms for the 10 Hz repetition rate. In conclusion the real deposition rate during the laser pulse is 10^{-2} nm/pulse, which is equivalent to 0.3×10^7 nm/s and it is much higher than any other PVD technique. Assuming a gas to solid phase transition which takes place in few ns, the equivalent cooling rate is immensely higher than in BMG growth; the self organization of Zr and Zr_2Cu is, therefore, of special importance. This is even more important if we compare their structure with that of glassy films grown by simultaneous Zr and Cu deposition (Fig. 4.14) with the same growth rate, proving that using the same lasing conditions may form glassy or nanostructured films depending on the nature of the target (homogeneous or sectored).

Acknowledgments The author would like to acknowledge Prof. C. Kosmidis for his long collaboration in developing the PLD system at the University of Ioannina and in laser research, Dr. D.C. Koutsogeorgis for the careful reading of the manuscript and for crucial comments, Prof. Ph. Komninou, Dr. G. Dimitrakopoulos, and Dr. Th. Kehagias for the TEM images presented in Figs. 4.10, 4.11, 4.14, 4.15, my former students G.M. Matenoglou, L.E. Koutsokeras, and H. Zoubos, and my fellow colleagues Ch.E. Lekka and G.A. Evangelakis for their research collaboration.

References

1. M. Zhang, M. Yudasaka, S. Iijima, Chem. Phys. Lett. **336**, 196 (2001)
2. Y. Zhang, H. Gu, S. Iijima, Appl. Phys. Lett. **73**, 3827 (1998)
3. E.H. Yang, S. Strauf, F. Fisher, D.S. Choi, Proc. SPIE **7318**, art. no. 731813 (2009)

4. M. Bystrzejewski, M.H. Rümmeli, H. Lange, A. Huczko, P. Baranowski, T. Gemming, T. Pichler, J. Nanosci. Nanotechnol. **8**, 6178 (2008)
5. A.J. Fetterman, Y. Raitses, M. Keidar, Carbon **46**, 1322 (2008)
6. J. Chae, X. Ho, J.A. Rogers, K. Jain, Appl. Phys. Lett. **92**, art. no. 173115 (2008)
7. J.I. Sohn, Y.-W. Ok, T.Y. Seong, S. Lee, J. Appl. Phys. **102**, art. no. 014301 (2007)
8. M. Kusaba, Y. Tsunawaki, Appl. Surf. Sci. **253**, 6330 (2007)
9. M. Chandran, K. Mohan Kant, N. Rama, M.S. Ramachandra Rao, MRS Symp. Proc. 951, 185 (2006)
10. S. Tamir, Y. Drezner, Appl. Surf. Sci. **252**, 4819 (2006)
11. G. Radhakrishnan, P.M. Adams, L.S. Bernstein, Thin Solid Films **515**, 1142 (2006)
12. M. Kusaba, Y. Tsunawaki, Thin Solid Films **506–507**, 255 (2006)
13. T. Ikuno, S.-I. Honda, K. Aoki, K. Oura, M. Katayama, Jpn. J. Appl. Phys. **A45**, 2872 (2006)
14. T. Ucda, H. Norimatsu, M.M.H. Bhuiyan, T. Ikegami, K. Ebihara, MRS Symp. Proc. **900**, 265 (2005)
15. N. Saurakhiya, Y.W. Zhu, F.C. Cheong, C.K. Ong, A.T.S. Wee, J.Y. Lin, C.H. Sow, Carbon **43**, 2128 (2005)
16. D.B. Geohegan, C.H. Schittenhelm, A.A. Puretzky, M.J. Lance, G.E. Jellison, P.F. Britt, Proc. SPIE **4977**, 658 (2003)
17. J.I. Sohn, C. Nam, S. Lee, Appl. Surf. Sci. **197–198**, 568 (2002)
18. M.-R. Chiang, K.-S. Liu, T.-S. Lai, C.-H. Tsai, H.-F. Cheng, I.-N. Lin, J. Vac. Sci. Technol. **B19**, 1034 (2001)
19. A.K. Sharma, R. Kalyanaraman, R.J. Narayan, S. Oktyabrsky, J. Narayan, Mater. Sci. Eng. **B79**, 123 (2001)
20. H. Wang, M. Chhowalla, N. Sano, S. Jia, G.A.J. Amaratunga, Nanotechnology **15**, 546 (2004)
21. P.M. Ossi, A. Miotello, J. non-Cryst. Solids **353**, 1860 (2007)
22. S. Ohmagari, T. Yoshitake, A. Nagano, R. Ohtani, H. Setoyama, E. Kobayashi, K. Nagayama, Diam. Relat. Mater. **19**, 911 (2010)
23. T. Yoshitake, S. Ohmagari, A. Nagano, S. Al-Riyami, R. Ohtani, H. Setoyama, E. Kobayashi, K. Nagayama, J. Nanomater., art. no. 876561 (2009)
24. E. Cappelli, C. Scilletta, S. Orlando, V. Valentini, M. Servidori, Appl. Surf. Sci. **255**, 5620 (2009)
25. E. Cappelli, C. Scilletta, G. Mattei, V. Valentini, S. Orlando, M. Servidori, Appl. Phys. **A93**, 751 (2008)
26. T. Yoshitake, A. Nagano, M. Itakura, N. Kuwano, T. Hara, K. Nagayama, Jpn. J. Appl. Phys. **L46**, 936 (2007)
27. J.D. Carey, S.J. Henley, Diam. Relat. Mater. **16**, 1782 (2007)
28. P. Patsalas, S. Kaziannis, C. Kosmidis, D. Papadimitriou, G. Abadias, G.A. Evangelakis, J. Appl. Phys. **101**, art. no. 124903 (2007)
29. L. Escobar-Alarcón, A. Arrieta, E. Camps, S. Romero, S. Muhl, M.A. Camacho-López, Diam. Relat. Mater. **16**, 1291 (2007)
30. S.J. Henley, J.D. Carey, S.R.P. Silva, G.M. Fuge, M.N.R. Ashfold, D. Anglos, Phys. Rev. **B72**, 1 (2005)
31. T. Katsuno, C. Godet, J.C. Orlianges, A.S. Loir, F. Garrelie, A. Catherinot, Appl. Phys. **A8**, 471 (2005)
32. G.M. Fuge, C.J. Rennick, S.R.J. Pearce, P.W. May, M.N.R. Ashfold, Diam. Relat. Mater. **12**, 1049 (2003)
33. T. Noguchi, K. Nagayama, Diam. Relat. Mater. **12**, 953 (2003)
34. E. Fogarassy, T. Szorényi, F. Antoni, G. Pirio, J. Olivier, P. Legagneux, P. Boher, Appl. Phys. **A76**, 15 (2003)
35. A.A. Voevodin, T.A. Fitz, J.J. Hu, J.S. Zabinski, J. Vac. Sci. Technol. **A20**, 1434 (2002)
36. P. Papakonstantinou, P. Lemoine, J. Phys. Cond. Mat. **13**, 2971 (2001)
37. K. Ebihara, T. Nakamiya, T. Ohshima, T. Ikegami, S.-I. Aoqui, Diam. Relat. Mater. **10**, 900 (2001)

38. V.I. Merkulov, D.H. Lowndes, L.R. Baylor, G.E. Jellison, A.A. Puretzky, D.B. Geohegan, SPIE Proc. **3618**, 495 (1999)
39. A.V. Rode, B. Luther-Davies, E.G. Gamaly, J. Appl. Phys. **85**, 4222 (1999)
40. I. Alexandrou, I. Zergioti, G.A.J. Amaratunga, M.J.F. Healy, C.J. Kiely, P. Hatto, M. Velegrakis, C. Fotakis, Mater. Lett. **39**, 97 (1999)
41. V.I. Merkulov, D.H. Lowndes, G.E. Jellison, A.A. Puretzky, D.B. Geohegan, Appl. Phys. Lett. **73**, 2591 (1998)
42. D.R. McKenzie, Rep. Progr. Phys. **59**, 1611 (1996)
43. M.P. Siegal, D.R. Tallant, L.J. Martinez-Miranda, J.C. Barbour, R.L. Simpson, D.L. Overmyer, Phys. Rev. **B61**, 10451 (2000)
44. T.A. Friedmann, J.P. Sullivan, J.A. Knapp, D.R. Tallant, D.M. Follstaedt, D.L. Medlin, P.B. Mirkarimi, Appl. Phys. Lett. **71**, 3820 (1997)
45. N. Sbaï-Benchikh, A. Zeinert, H. Caillièrez, C. Donnet, Diam. Relat. Mater. **18**, 1085 (2009)
46. S. Bhattacharyya, S.J. Henley, D. Lock, N.P. Blanchard, S.R.P. Silva, Appl. Phys. Lett. **89**, art. no. 022113 (2006)
47. S.J. Henley, N.E.P. Woolger, J.D. Carey, S.R.P. Silva, G.M. Fuge, M.N.R. Ashfold, Mater. Res. Soc. Symp. Proc. **876**, 256 (2005)
48. A.V. Rode, E.G. Gamaly, B. Luther-Davies, Appl. Phys. **A70**, 135 (2000)
49. F. Pinakidou, M. Katsikini, M. Zougrou, G.M. Matenoglou, P. Patsalas, E.C. Paloura, Thin Solid Films, *in press* (2011)
50. A.A. Voevodin, S.V. Prasad, J.S. Zabinski, J. Appl. Phys. **82**, 855 (1997)
51. A. Lotsari, G.P. Dimitrakopulos, Th. Kehagias, P. Kavouras, H. Zoubos, L.E. Koutsokeras, P. Patsalas, Ph. Komninou, Surf. Coat. Technol. **204**, 1937 (2010)
52. F. Pinakidou, E.C. Paloura, G.M. Matenoglou, P. Patsalas, Surf. Coat. Technol. **204**, 1933 (2010)
53. G.M. Matenoglou, H. Zoubos, A. Lotsari, Ch.E. Lekka, Ph. Komninou, G.P. Dimitrakopulos, C. Kosmidis, G.A. Evangelakis, P. Patsalas, Thin Solid Films **518**, 1508 (2009)
54. G.M. Matenoglou, G.A. Evangelakis, C. Kosmidis, S. Foulias, D. Papadimitriou, P. Patsalas, Appl. Surf. Sci. **253**, 8155 (2007)
55. B. Zhuo, Y. Li, S. Teng, A. Yang, Appl. Surf. Sci. **256**, 3305 (2010)
56. A.V. Zenkevich, Y.Y. Lebedinskii, A.A. Timofeyev, I.A. Isayev, V.N. Tronin, Appl. Surf. Sci. **255**, 5355 (2009)
57. C.N. Hunter, M.H. Check, J.E. Bultman, A.A. Voevodin, Surf. Coat. Technol. **203**, 300 (2008)
58. M.R.S. Castro, E.D. Sam, M. Veith, P.W. Oliveira, Nanotechnology **19**, art. no. 105704 (2008)
59. A. Crunteanu, F. Dumas-Bouchiat, C. Champeaux, A. Catherinot, P. Blondy, Thin Solid Films **515**, 6324 (2007)
60. J.J. Lin, T. Zhang, P. Lee, S.V. Springham, T.L. Tan, R.S. Rawat, T. White, R. Ramanujan, J. Guo, Appl. Phys. Lett. **91**, art. no. 063120 (2007)
61. E. György, G. Sauthier, A. Figueras, A. Giannoudakos, M. Kompitsas, I.N. Mihailescu, J. Appl. Phys. **100**, art. no. 114302 (2006)
62. R. Serna, A. Suárez-García, C.N. Afonso, D. Babonneau, Nanotechnology **17**, 4588 (2006)
63. G. Compagnini, Appl. Surf. Sci. **226**, 216 (2004)
64. W. Wang, G. Yang, W. Wu, Z. Chen, J. Appl. Phys. **94**, 6837 (2003)
65. T. Sasaki, K.M. Beck, N. Koshizakai, Appl. Surf. Sci. **197–198**, 619 (2002)
66. R. Serna, D. Babonneau, A. Suárez-García, C.N. Afonso, E. Fonda, A. Traverse, A. Naudon, D.E. Hole, Phys. Rev. **B66**, art. no. 205402 (2002)
67. C.N. Afonso, R. Serna, J.M. Ballesteros, A.K. Petford-Long, R.C. Doole, Appl. Surf. Sci. **127–129**, 339 (1998)
68. H. Wang, X. Zhang, A. Gupta, A. Tiwari, J. Narayan, Appl. Phys. Lett. **83**, 3072 (2003)
69. H. Yang, H. Wang, B. Maiorov, J. Lee, D. Talbayev, M.J. Hinton, D.M. Feldmann, J.L. MacManus-Driscoll, A.J. Taylor, L. Civale, T.R. Lemberger, Q.X. Jia, J. Appl. Phys. **106**, art. no. 093914 (2009)
70. Z. Bi, J.H. Lee, H. Yang, Q. Jia, J.L. MacManus-Driscoll, H. Wang, J. Appl. Phys. **106**, art. no. 094309 (2009)

71. J. Barbosa, B.G. Almeida, J.A. Mendes, A.G. Rolo, J.P. Araújo, J.B. Sousa, J. Appl. Phys. **101**, art. no. 09101 (2007)
72. B.S. Kang, H. Wang, J.L. MacManus-Driscoll, Y. Li, Q.X. Jia, I. Mihut, J.B. Betts, Appl. Phys. Lett. **88**, art. no. 192514 (2006)
73. J.-P. Zhou, H. He, Z. Shi, C.-W. Nan, Appl. Phys. Lett. **88**, art. no. 013111 (2006)
74. H. Kadokura, A. Ito, T. Kimura, T. Goto, Surf. Coat. Technol. **204**, 2302 (2010)
75. I. Shishkovsky, Y. Morozov, I. Yadroitsev, I. Smurov, Appl. Surf. Sci. **255**, 9847 (2009)
76. Z. Liu, D.J. Styers-Barnett, A.A. Puretzky, C.M. Rouleau, D. Yuan, I.N. Ivanov, K. Xiao, J. Liu, D.B. Geohegan, Appl. Phys. **A93**, 987 (2008)
77. M. Kwoka, L. Ottaviano, M. Passacantando, G. Czempik, S. Santucci, J. Szuber, Appl. Surf. Sci. **254**, 8089 (2008)
78. P. Heszler, L. Landström, C.G. Granqvist, Appl. Surf. Sci. **253**, 8292 (2007)
79. Y. Maezono, K. Toshikawa, K. Kurosawa, K. Amari, S. Ishimura, M. Katto, A. Yokotani, Jpn. J. Appl. Phys. **46**, 3534 (2007)
80. T. Goto, Surf. Coat. Technol. **198**, 367 (2005)
81. B. Wu, P.I. Cohen, L.C. Feldman, Z. Zhang, Appl. Phys. Lett. **84**, 2175 (2004)
82. A. Santoni, J. Lancok, S. Loreti, I. Menicucci, C. Minarini, F. Fabbri, D. Della Sala, J. Cryst. Growth **258**, 272 (2003)
83. K.H. Kwok, W.K.S. Chiu, Carbon **41**, 2307 (2003)
84. Y. Damlag, A. Goossens, I. Colbeck, J. Schoonman, Adv. Mater. **15**, 125 (2003)
85. D. Tonneau, F. Thuron, A. Correia, J.E. Bouree, Y. Pauleau, Jpn. J. Appl. Phys. **37**, 4954 (1998)
86. D.C. Koutsogeorgis, W.M. Cranton, R.M. Ranson, C.B. Thomas, J. All. Comp. **483**, 526 (2009)
87. S.V. Yap, R.M. Ranson, W.M. Cranton, D.C. Koutsogeorgis, Appl. Opt. **47**, 4895 (2008)
88. D.C. Koutsogeorgis, E.A. Mastio, W.M. Cranton, C.B. Thomas, Thin Solid Films **383**, 31 (2001)
89. Y.-T. Cheng, R.-H. Uang, Y.-M. Wang, K.-C. Chiou, T.-M. Lee, Microelectronic Eng. **86**, 865 (2009)
90. P.I. Gaiduk, S.L. Prakopyeu, V.A. Zajkov, G.D. Ivlev, E.I. Gatskevich, Physica **B404**, 4708 (2010)
91. G.A. Cirino, R.D. Mansano, P. Verdonck, R.G. Jasinevicius, L.G. Neto, Surf. Coat. Technol. **204**, 2966 (2010)
92. D. Cammilleri, F. Fossard, M. Halbwax, C.T. Manh, N. Yam, D. Débarre, J. Boulmer, D. Bouchier, Thin Solid Films **517**, 327 (2008)
93. M. Shimizu, Y. Shimotsuma, M. Sakakura, T. Yuasa, H. Homma, Y. Minowa, K. Tanaka, K. Miura, K. Hirao, Optics Exp. **17**, 46 (2009)
94. Y. Lin, A. Harb, D. Rodriguez, K. Lozano, D. Xu, K.P. Chen, J. Appl. Phys. **104**, 113111 (2008)
95. P. Olivero, S. Calusi, L. Giuntini, S. Lagomarsino, A. Lo Giudice, M. Massi, S. Sciortino, M. Vannoni, E. Vittone, Diam. Relat. Mater. **19**, 428 (2010)
96. J. Preclíková, A. Kromka, B. Rezek, P. Malý, Opt. Lett. **35**, 577 (2010)
97. M. Neff, T.V. Kononenko, S.M. Pimenov, V. Romano, W. Lüthy, V.I. Konov, Appl. Phys. **A97**, 543 (2009)
98. T. Okuchi, H. Ohfuji, S. Odake, H. Kagi, S. Nagatomo, M. Sugata, H. Sumiya, Appl. Phys. **A96**, 833 (2009)
99. J. Kanasaki, E. Inami, K. Tanimura, H. Ohnishi, K. Nasu, Phys. Rev. Lett. **102**, 087402 (2009)
100. T.V. Kononenko, M. Meier, M.S. Komlenok, S.M. Pimenov, V. Romano, V.P. Pashinin, V.I. Konov, Appl. Phys. **A90**, 645 (2008)
101. V.S. Pavelyev, S.A. Borodin, N.L. Kazanskiy, G.F. Kostyuk, A.V. Volkov, Optics Laser Technol. **39**, 1234 (2007)
102. G. Dumitru, V. Romano, H.P. Weber, S. Pimenov, T. Kononenko, M. Sentis, J. Hermann, S. Bruneau, Appl. Surf. Sci. **222**, 226 (2004)

103. D. Vouagner, Cs. Beleznai, J.P. Girardeau-Montaut, C. Templier, H. Gonnord, Appl. Surf. Sci. **154–155**, 201 (2000)
104. Z. Yan, R. Bao, D.B. Chrisey, Nanotechnology **21**, art. no. 145609 (2010)
105. N.G. Semaltianos, S. Logothetidis, N. Hastas, W. Perrie, S. Romani, R.J. Potter, G. Dearden, K.G. Watkins, P. French, M. Sharp, Chem. Phys. Lett. **484**, 283 (2010)
106. D.C. Schinca, L.B. Scaffardi, F.A. Videla, G.A. Torchia, P. Moreno, L. Roso, J. Phys. **D42**, art. no. 215102 (2009)
107. N. Bärsch, J. Jakobi, S. Weiler, S. Barcikowski, Nanotechnology **20**, art. no. 445603 (2009)
108. S.J. Henley, S. Mollah, C.E. Giusca, S.R.P. Silva, J. Appl. Phys. **106**, art. no. 064309 (2009)
109. T. Tsuji, M. Nakanishi, T. Mizuki, M. Tsuji, T. Doi, T. Yahiro, J. Yamaki, Appl. Surf. Sci. **255**, 9626 (2009)
110. E. Giorgetti, A. Giusti, F. Giammanco, P. Marsili, S. Laza, Molecules **14**, 3731 (2009)
111. P. Liu, W. Cai, M. Fang, Z. Li, H. Zeng, J. Hu, X. Luo, W. Jing, Nanotechnology **20**, art. no. 285707 (2009)
112. K. Kawaguchi, R. Wu, Y. Ishikawa, T. Sasaki, N. Koshizaki, J. Nanosci. Nanotechnol. **9**, 1454 (2009)
113. A. Spangenberg, R. Métivier, J. Gonzalez, K. Nakatani, P. Yu, M. Ciraud, A. Léaustic, R. Guillot, T. Uwada, T. Asahi, Adv. Mater. **21**, 309 (2009)
114. K. Oguri, Y. Okano, T. Nishikawa, H. Nakano, Phys. Rev. Lett. **99**, art. no. 165003 (2007)
115. S. Barcikowski, A. Meńndez-Manjón, B. Chichkov, M. Brikas, G. Račiukaitis, Appl. Phys. Lett. **91**, art. no. 083113 (2007)
116. H. Zeng, W. Cai, J. Hu, G. Duan, P. Liu, Y. Li, Appl. Phys. Lett. **88**, art. no. 171910 (2006)
117. M.N.R. Ashfold, F. Claeyssens, G.M. Fuge, S.J. Henley, Chem. Soc. Rev. **33**, 23 (2004)
118. F. Xiong, Y.Y. Wang, R.P.H. Chang, Phys. Rev. **B48**, 8016 (1993)
119. P.M. Ossi, C.E. Bottani, A. Miotello, Thin Solid Films **482**, 2 (2005)
120. J. Robertson, Mater. Sci. Eng. **R37**, 129 (2002)
121. F. Balon, V. Stolojan, S.R.P. Silva, M. Michalka, A. Kromka, Vacuum **80**, 163 (2005)
122. C.-R. Wang, R.-B. Huang, Z.-Y. Liu, and L. S. Zheng, Chem. Phys. Lett. **227**, 109 (1994)
123. D.L. Pappas, K.L. Saenger, J. Bruley, W. Krakow, J.J. Cuomo, T. Gu, R.W. Collins, J. Appl. Phys. **71**, 5675 (1992)
124. A.C. Ferrari, A. Libassi, B.K. Tanner, V. Stolojan, J. Yuan, L.M. Brown, S.E. Rodil, B. Kleinsorge, J. Robertson, Phys. Rev. **B62**, 11089 (2000)
125. Y.-K. Choi, H.-S. Im, K.-W. Jung, Int. J. Mass. Spectrom. **189**, 115 (1999)
126. K. Yamamoto, Y. Koga, S. Fujiwara, F. Kokai, R.B. Heimann, Appl. Phys. **A66**, 115 (1998)
127. K. Yamamoto, Y. Koga, S. Fujiwara, F. Kokai, Jpn. J. Appl. Phys. **L36**, 1333 (1997)
128. S. Bakalova, A. Szekeres, A. Cziraki, C.P. Lungu, S. Grigorescu, G. Socol, E. Axente, I.N. Mihailescu, Appl. Surf. Sci. **253**, 8215 (2007)
129. B. Howe, J. Bareño, M. Sardela, J.G. Wen, J.E. Greene, L. Hultman, A.A. Voevodin, I. Petrov, Surf. Coat. Technol. **202**, 809 (2007)
130. L. Chen, Y. Du, P.H. Mayrhofer, S.Q. Wang, J. Li, Surf. Coat. Technol. **203**, 5158 (2008)
131. D. Rafaja, C. Wüstefeld, M. Dopita, V. Klemm, D. Heger, G. Schreiber, M. Šíma, Surf. Coat. Technol. **203**, 572 (2008)
132. S.H. Sheng, R.F. Zhang, S. Veprek, Acta Materialia **56**, 968 (2008)
133. M. Odén, L. Rogström, A. Knutsson, M.R. Terner, P. Hedström, J. Almer, J. Ilavsky, Appl. Phys. Lett. **94**, 053114 (2009)
134. J.C. Oliveira, A. Cavaleiro, M.T. Vieira, Surf. Coat. Technol. **151–152**, 466 (2002)
135. Z.G. Wu, G.A. Zhang, M.X. Wang, X.Y. Fan, P.X. Yan, T. Xu, Appl. Surf. Sci. **253**, 2733 (2006)
136. J. Toudert, D. Babonneau, S. Camelio, T. Girardeau, F. Yubero, J.P. Espinós, A.R. Gonzalez-Elipe, J. Phys. **D40**, 4614 (2007)
137. D. Babonneau, J. Toudert, S. Camelio, F. Pailloux, T. Cabioch, T. Girardeau, Surf. Coat. Technol. **200**, 6251 (2006)
138. A.J. McAlister, Bull. Alloy Phase Diagrams **8**, 526 (1987)
139. H. Okamoto, J. Phase Equil. Diff. **26**, 391 (2005)

140. Y. Zhang, J.K. Liang, J.-B. Li, Q.L. Liu, Y.G. Xiao, Q. Zhang, B.J. Sun, G.H. Rao, J. Alloys. Comp. **429**, 184 (2007)
141. R.J. Narayan, Appl. Surf. Sci. **245**, 420 (2005)
142. R. Serna, A. Suarez-Garcia, C.N. Afonso, D. Babonneau, Nanotechnology **17**, 4588 (2006)
143. W. Wang, G. Yang, W. Wu, Z. Chen, J. Appl. Phys. **94**, 6837 (2003)
144. H. Wang, X. Zhang, A. Gupta, A. Tiwari, J. Narayan, Appl. Phys. Lett. **83**, 3072 (2003)
145. L.E. Koutsokeras, G. Abadias, Ch.E. Lekka, G.M. Matenoglou, D.F. Anagnostopoulos, G.A. Evangelakis, P. Patsalas, Appl. Phys. Lett. **93**, 011904 (2008)
146. Ch. E. Lekka, P. Patsalas, Ph. Komninou, G.A. Evangelakis, J. Appl. Phys., *in press* (2011)
147. G.A. Almyras, G.M. Matenoglou, Ph. Komninou, C. Kosmidis, P. Patsalas, G.A. Evangelakis, J. Appl. Phys. **107**, 8 (2010)
148. W.L. Johnson, Mater. Res. Soc. Bull. **24**, 42 (1999)
149. A. Inoue, Acta Mater. **48**, 279 (2000)
150. J. Schroers, W.L. Johnson, Phys. Rev. Lett. **93**, 255506 (2004)
151. Y. Zhang, W.H. Wang, A.L. Greer, Nat. Mater. **5**, 857 (2006)
152. H.W. Sheng, H.Z. Liu, Y.Q. Cheng, J. Wen, P.L. Lee, W.K. Luo, S.D. Shastri, E. Ma, Nat. Mater. **6**, 192 (2007)
153. F. Spaepen, Nature **408**, 781 (2000)
154. A.R. Yavari, Nature **439**, 405 (2006)
155. C.A. Schuh, A.C. Lund, Nat. Mater. **2**, 449 (2003)
156. Ch.E. Lekka, A. Ibenskas, A.R. Yavari, G.A. Evangelakis, Appl. Phys. Lett. **91**, 214103 (2007)

Chapter 5
High Efficiency Multijunction Solar Cells with Finely-Tuned Quantum Wells

Argyrios C. Varonides

The field of high efficiency (inorganic) photovoltaics (PV) is rapidly maturing in both efficiency goals and cover all cost reduction of fabrication. On one hand, know-how from space industry in new solar cell design configurations and on the other, fabrication cost reduction challenges for terrestrial uses of solar energy, have paved the way to a new generation of PV devices, capable of capturing most of the solar spectrum. For quite a while now, the goal of inorganic solar cell design has been the total (if possible) capture-absorption of the solar spectrum from a single solar cell, designed in such a way that a multiple of incident wavelengths could be simultaneously absorbed. Multi-absorption in device physics indicates parallel existence of different materials that absorb solar photons of different energies. Bulk solid state devices absorb at specific energy thresholds, depending on their respective energy gap (E_G). More than one energy gaps would on principle offer new ways of photon absorption: if such a structure could be fabricated, two or more groups of photons could be absorbed simultaneously. The point became then what lattice-matched semiconductor materials could offer such multiple levels of absorption without much recombination losses. It was soon realized that such layer multiplicity combined with quantum size effects could lead to higher efficiency collection of photo-excited carriers. At the moment, the main reason that slows down quantum effect solar cell production is high fabrication cost, since it involves primarily expensive methods of multilayer growth. Existing multi-layer cells are fabricated in the bulk, with three (mostly) layers of lattice-matched and non-lattice-matched (pseudo-morphic) semiconductor materials (GaInP/InGaN etc), where photo-carrier collection occurs in the bulk of the base (coming from the emitter which lies right under the window layer). These carriers are given excess to conduction via tunnel junction (grown between at each interface and connecting the layers in series). This basic idea of a design has proven very successful in recent years, leading to

A.C. Varonides (✉)
Physics and EE Department, University of Scranton, Scranton, PA, USA
e-mail: varonides@scranton.edu

S. Logothetidis (ed.), *Nanostructured Materials and Their Applications*, NanoScience and Technology, DOI 10.1007/978-3-642-22227-6_5,

solar cells of efficiency levels well above 30% (Fraunhofer Institute's multi-gap solar cell at 40.8%, and NREL's device at 40.2% respectively). Successful alloys have demonstrated high performance, such as $In_xGa_{1-x}P$ alloys (x (%) of gallium phosphide and (1 − x) (%) of indium phosphide). Other successful candidates, in current use and perpetual cell design consideration, are the lattice-matched GaAs/AlGaAs and InP/GaAs pairs or AlAs/GaAs/GaAs triple layers and alloys, which are heavily used in both solar and the electronics industry. In this chapter, combinations of lattice-matched layers are considered in a current high-efficiency solar cell design. In spite the fabrication costs, these multilayered solar cells are here to stay: international collaborative research and development efforts are currently leading to high and very-high efficiency solar cells of tiny size, embedded in sophisticated Fresnel-optics tracking systems with great expectations (concentrated photovoltaics or CPV).

5.1 What is a Solar Cell?

A p-n junction is necessary for a solar cell to operate under illumination. Basically, solar photons of a specific range (either from the visible or the infrared) may be absorbed by semiconductor layers involved in the cell structure. The energy band of a semiconductor includes the valence and conduction bands, separated by a forbidden gap of energy. The latter is characteristic of the medium involved (e.g., for silicon, the gap energy is 1.12 and for germanium 0.67 eV respectively). The cell structure is fundamentally a p-n junction: given a semiconductor, there has to be a way to extract free carriers by means of illumination. Incident photons are supposed to be absorbed by the medium at energy values very near the gap energy of the medium. For free carriers to be collected as measurable external current, a specific design must exist: somehow, an electrostatic field that could sweep out the carriers from the region of their generation, before these carriers recombine (lost or return to the valence band). Non-zero fields can be formed via p- and n-regions in direct contact. Therefore, a typical bulk cell is implanted with two p and n regions in contact, thus forming a p-n junction. As p- and n- carriers form space charge regions across the junction, a depletion region is formed with non zero electric field which is maximum at the interface and decrease down to zero at the edges of the two regions. When a carrier finds itself in the depletion region, it eventually accelerates along or against the field (depending on its charge). The latter process is the key operation for all solid state devices: their basic function is based on the existence of one or more p-n junctions (optical devices are based on p-n junctions as the main regions where electrons and holes can essentially separate from each other to form photo-currents). Typically, in an illuminated p-n junction, photons are absorbed and electron-hole pairs are generated. These carriers eventually diffuse in opposite directions (separated by the existing electrostatic field at the junction), and within their respective diffusion lengths. Electrons at the p-side slide down the junction potential and holes get to the opposite directions. Under open-circuit conditions, the voltage across the cell is given by the following formula:

$$V_{oc} = kT \ln\left(1 + \frac{I_L}{I_o}\right) \tag{5.1}$$

Where k is Boltzmann's constant, T(°K) is temperature, I_L is the light-generated current, and I_o is the p-n junction's reverse saturation current. On the other hand, when it comes to the experimental characterization of the cells, a quantity called internal quantum efficiency is useful: the number of excited electrons, for each absorbed photon.

5.2 Photo-Currents

Solar photons strike the surface of the cell at all wavelengths, from below the edge of the visible to near far infrared (IR). Just above the earth's atmosphere, the solar spectrum is similar to 6,000°K black body (AM0 conditions). At sea level, the atmospheric absorption modifies the solar irradiance ($\mathrm{mW/cm^2}$) to about $100\,\mathrm{mW/cm^2}$ (AM1.5 conditions). Photons of energy below the energy gap do not get absorbed by the material; photons with energies at the gap edge and above have a finite chance of being absorbed. Fundamentally, each photon will generate one hole-electron pair. This means that the number of photo-carriers is equal to the number of incident photons that are absorbed. Far from the junction, the electric fields are weak and carrier transport occurs by diffusion. In the n-region, holes diffuse to the junction, where they are collected and dragged by the field to the other side. Similarly, electrons in the p-region diffuse to the edge of the depletion region and get swept to the other side by means of the field in the depletion region. Fundamentally, solar cell modeling correlates incident solar photon flux F_{ph} (# $\mathrm{cm^{-2}\,s^{1}}$) with generation and recombination carrier rates in the interior of the device. Photo-generated concentrations of diffusing carriers are typically found (modeled) through the diffusion equation (under appropriate boundary conditions):

$$D_p \frac{d^2 p_n}{dx^2} - \frac{p_n - p_{no}}{\tau_p} + \alpha(1 - R) F_{ph} e^{-\alpha(x+d)} = 0 \tag{5.2}$$

Where p_n and p_{no} are carrier concentration after and before illumination (note excess carrier concentration is ($\delta p = p_n - p_{no}$), D_p is carrier diffusion coefficient, α is the material absorption coefficient, R is the reflection coefficient of the n-layer, and where x represents one-dimensional diffusion of carriers through the p-n junction. Typical next step after solving for p_n or n_p is diffusion current evaluation via standard techniques and based on the following:

$$J_p = -qD_p \left(\frac{dp_n}{dx}\right)_{x=x(j)} \tag{5.3}$$

$$J_n = qD_p \left(\frac{dn_p}{dx}\right)_{x=x(j)} \tag{5.4}$$

Photon-collection efficiency is usually defined as:

$$\eta_{\mathrm{col}} = \frac{J_p + J_n}{qF_{ph}} \tag{5.5}$$

Boundary conditions include continuity of carrier concentrations at the junction $x(j)$, and the dependence of the first derivative of carrier concentration on recombination velocity s_p, at the edge of the window layer as shown in the figure below:

$$\left(\frac{dp}{dx}\right)_{x=-d} = \frac{s_p}{D_p}(p(-d) - p_{no}) \tag{5.6}$$

The figure below indicates a generally accepted modeling geometry for a p-n junction solar cell.

5.3 Solution of the Diffusion Equation: n-Region

Minority holes are expected to be generated in the window layer (x from $-d$ to 0)

$$p_n(x) - p_{no} = A\cosh(x/L_p) + B\sinh(x/L_n) - \frac{\alpha F_{ph}(1-R)(L_p^2/D_p)}{(\alpha L_p)^2 - 1} e^{-\alpha(x+d)} \tag{5.7}$$

Based on the boundary conditions as mentioned above:

$$A = \frac{\alpha F_{ph}(1-R)(L_p^2/D_p)}{(\alpha L_p)^2 - 1} e^{-\alpha d} \tag{5.8}$$

$$B = \frac{\alpha F_{ph}(1-R)(L_p^2/D_p)}{(\alpha L_p)^2 - 1}\tanh(d/L_p) + \frac{\alpha^3 F_{ph}(1-R)(L_p^3/D_p)}{\cosh(d/L_p)(\alpha^2 L_p^2 - 1)} + \frac{s_p L_p}{D_p}\frac{[p(-d) - p_{no}]}{\cosh(d/L_p)} \tag{5.9}$$

Substitution of A, B constants provides a complete solution for minority holes in the n-region of the cell. Total electric current J_p coming from the n-region is naturally following from the diffusion current formula (see above). Specifically, maximum hole-current from the n-region is:

$$J(x=0) = -qD_pK \times \left[\frac{1}{L_p}\left[\tanh\left(\frac{d}{L_p}\right) + \frac{\alpha L_p}{\cosh(d/L_p)} + \frac{s_p L_p}{D_{pK}} \times \frac{p(-d) - p_{no}}{\cosh(d/L_p)}\right] + \alpha e^{-\alpha d}\right] \tag{5.10}$$

Where

$$K = \frac{\alpha F_{ph}(1-R)(L_p^2/D_p)}{(\alpha L_p)^2 - 1} \tag{5.11}$$

5.4 Solution of the Diffusion Equation: P-Region

Minority electrons develop in the p-region under the following boundary conditions:

$$[n_p - n_{po}]_{x=W} = 0 \tag{5.12}$$

And

$$-D_n\left(\frac{dn}{dx}\right)_{x=L+W} = s_n(n_p - n_{po}) \tag{5.13}$$

The diffusion equation reads as follows:

$$D_n\frac{d^2n_p}{dx^2} - \frac{n_p - n_{po}}{\tau_p} + \alpha(1-R)F_{ph}e^{-\alpha(x+d)} = 0 \tag{5.14}$$

Solution of (5.14) is of similar kind with (5.7) along with boundary conditions (5.13):

$$n - n_{po} = A\cosh(x/L_n) + B\sinh(x/L_n) - \frac{\alpha F_{ph}(1-R)(L_n^2/D_n)}{(\alpha L_n)^2 - 1}e^{\alpha(x+d)} \tag{5.15}$$

With A, as in (5.8):

$$A = \frac{\alpha F_{ph}(1-R)(L_n^2/D_n)}{(\alpha L_n)^2 - 1}e^{-\alpha d} \tag{5.16}$$

And B as in (5.9):

$$B = -\frac{s_n L_n}{D_n} \times \frac{n_p({}_nL + w) - n_{po}}{} \times \cosh\left(\frac{L_n + w}{L_n}\right) - \frac{\alpha F_{ph}(1-R)(L_n^2/D_n)}{(aL_n)^2 - 1}$$

$$\times \tanh\left(\frac{L_n + w}{L_n}\right) - \frac{\alpha^2 F_{ph}(1-R)(L_n^3/D_n)}{((\alpha L_n)^2 - 1)\cosh(\frac{L_n + w}{L_n})} \times \exp(-(d + L + w)) \tag{5.17}$$

5.5 Total Electron and Hole Currents

Electron current:

$$J_n = \frac{qD_n}{L_n} \times \frac{\alpha F_{ph}(1-R)\tau_n}{(\alpha L_n)^2 - 1}$$

$$\times \left\langle \sinh(W/L_n) - \left[\frac{s_n L_n}{D_n} \times \frac{n_p(w + L) - n_{po}}{\cosh\left(\frac{L+w}{L_n}\right)} \times \frac{(\alpha L_n)^2 - 1}{\alpha F_{ph}(1-R)\tau_n}\right.\right.$$

$$\left.\left. + \tanh\left(\frac{W}{L_n}\right) + \frac{e^{-\alpha(d+L+W)}}{\cosh\left(1 + \frac{W}{L_n}\right)}\right] \times \cosh\left(\frac{W}{L_n}\right) + \alpha L_n e^{-\alpha(W+d)}\right\rangle \tag{5.18}$$

Where $\tau_n = L_n^2/D_n$
Hole-current:

$$J_p = -\frac{qD_p}{L_p}\frac{\alpha F_{ph}(1-R)\tau_p}{(\alpha L_n)^2 - 1}$$

$$\times \left\{\tanh\left(\frac{d}{L_p}\right) + \frac{\alpha L_p}{\cosh\left(\frac{d}{L_p}\right)}\right.$$

$$\left. + \frac{s_p L_p}{D_p}\frac{(\alpha L_p)^2 - 1}{\alpha F_{ph}(1-R)\tau_p}\frac{p(-d) - p_{no}}{\cosh\left(\frac{d}{L_p}\right)} + \alpha L_p e^{-\alpha d}\right\} \tag{5.19}$$

5.6 P-I-N Geometries of Solar Cells

P-n junctions with long intrinsic (less impurity scattering) regions provide a wide range for photo-electrons to escape the depletion region (before recombination) and to be collected at the load. This advantage combined with the idea of using a heterojunction p-n interface instead of a bulk counterpart has led to many exciting cell designs. In the figure below, one may observe the discrete energy levels due to two semiconductors of different band gaps.

Candidate materials in such cases could be AlAs and GaAs for wide and narrow band gaps, respectively. Coupled with the idea of capturing a wider range of photons, the intrinsic region can be replaced by a sequence of layers with different band gaps (a superlattice) as shown in Fig. 5.1, below:

Photo-Carriers are trapped in quantum wells, and subsequently (a) tunnel through layers and/or (b) escape thermionically from the quantum wells. Initial selection of width can lead to specific eigen-energy level formation, as shown on the above. The ground and second (at the most) states provide carrier populations near 10^{12} cm^{-2}. If the second energy level is near the edge of the wells, it will eventually lead to smooth electron hopping, while tunneling might dominate at the ground level. If the thickness of the wells is selected such that only one solution exists in the wells, then carriers either escape thermally or tunnel through thin barriers. Thermal escape however will be dominant when thick barriers are selected (which is the case of Fig. 5.2).

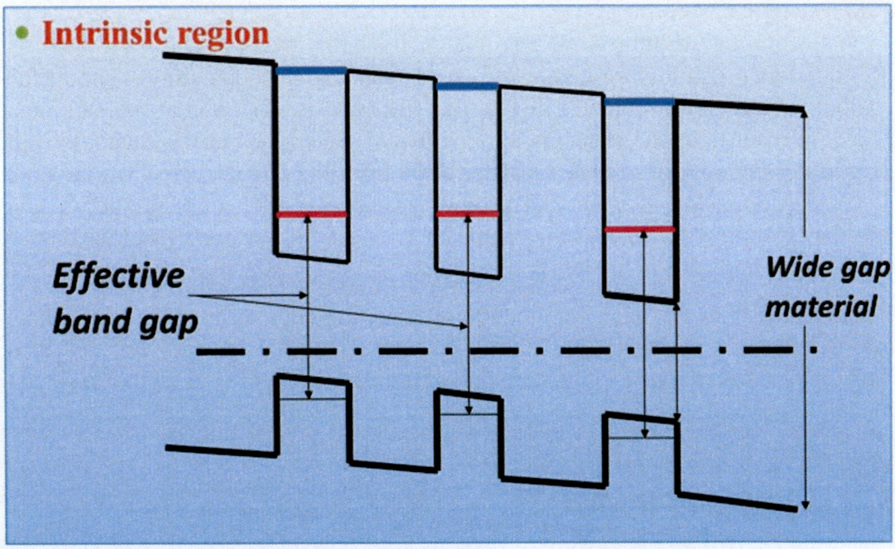

Fig. 5.1 Detail from a superlattice structure (typically GaAs/alloy and GaAs/Ge (as proposed in this study). The dashed line represents the Fermi level at thermal equilibrium. The optical gap can be tuned to desired energy values

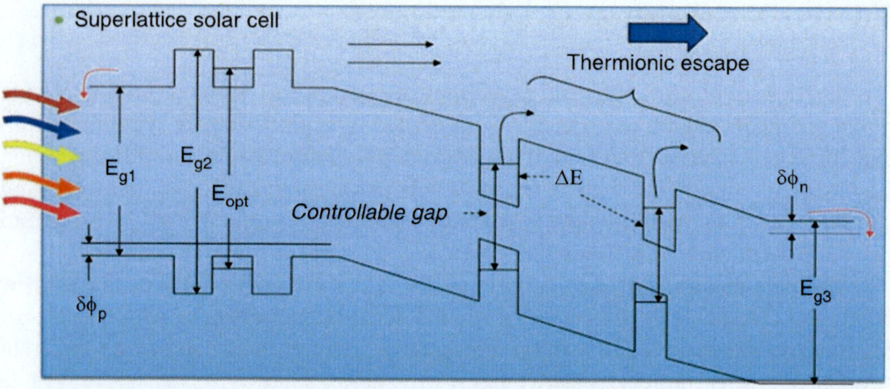

Fig. 5.2 An improved solar cell design with more than two layers. Note the existence of quantum traps in the intrinsic region, and the quantum well in the p-region (far left)

As seen from Fig. 5.2, a number of issues are of concern:

1. Note the double-barrier (DB) structure in the p-region: it may be a set of alternate GaAs/AlAs/GaAs layers, where the DB structure will provide excess photocarriers at desired (long) solar wavelengths.
2. Note the proposed intrinsic region: a sequence of multiple quantum wells (MQWs) is proposed, so that desired incident solar wavelengths can be absorbed as well (especially in the 1 eV range).
3. The last layer is proposed to be either GaAs or Ge (0.66 eV hence long wavelength absorption).
4. Visible light absorption made possible via the AlAs layers (far left).
5. Tunneling possibility through energy eigen-states in the quantum wells: note the double well (DW) structure in Fig. 5.2, where transmission is at its maximum at peak values of the transmission coefficient. Selecting barrier thickness well thickness, may lead to peak current values from the AlAs layer down the bulk GaAs region. Note also that DW geometries may play a key role in multijunction solar cells: they provide a series connection layer to lower portions of the device (MJ cells and cells in tandem).

Based on the above points, combination of tunneling and thermionic currents may lead to wide solar spectrum absorption, depending on the design adopted. Note also the tunnel-diode structure on the left side of the device, with sequential or resonant tunneling possibilities.

5.7 A Proposed Device

Very recently, we have proposed an undoped GaAs-Ge MQW in a standard p-i-n design, namely, p-intrinsic (MQW)-n geometry that includes lattice-matched GaAs and Ge layers in the intrinsic region of the PV device. This formation offers the advantage at 1 eV absorption in the III–V realm, without compromises in device transport properties, such as mobility or conductivity. GaAs-based layers ensure (a) high mobility and absorption values (via GaAs and *thin* Ge layers) and (b) fine-tuning of the *optical gap* with solar photon wavelength.

The proposed design is a part of a more complex structure that includes the top and bottom cells in tandem. Calculations regarding both parts of the device are showing results as follows: (a) Top sub-cell (AlAs-AlGaAs-GaAs): in excess of 21% (Varonides et al., WREC 10, Glasgow 2008) conversion efficiency predicted for a 1 cm^2 cross section lattice-matched AlAs/AlGaAs/GaAs unit in the visible (AM1.5, one sun). Modeling is based on bulk layer calculations for n^+ (window)–n–p triple junction solar cell. As second choice for a top layer, we include a 30% metamorphic triple-junction InP-GaAs-GaAs solar cell. (b) Bottom cell: eigen-energy fine tuning calculations lead to carrier confinement and subsequent carrier escape with thermionic emission and nearest neighbor hopping (NNH) conduction from site to site. Carrier escape from the quantum wells is thought to be a thermionic process, along with recombination losses. Incident IR photons are expected to be absorbed in the MQW area of the structure, while recombination losses are taken into account. Projected excess carriers (electrons) are of the order of 10^{12}–10^{13} cm^{-2} per ground eigen-state. Thermionic current density values are then projected in the order of 30 mA/cm^2 and open-circuit voltage values above 1 V, at one sun. Overall (composite device) collection efficiency values are initially projected well in excess of 40%, which is a key threshold target for current high efficiency PV. Photon shadowing issues can eventually be overcome with appropriate optical design (for instance, cell placed at the focal plane of a parabolic or concave mirror, where both sides are illuminated simultaneously).

Total current density is dominated by the lowest of the two sub-cell currents, and open-circuit voltage values are the sum of the two sub-cell V_{oc} values. Total current from the bottom cell is the sum of thermionic and nearest neighbor hopping currents (NNH, Varonides, and Spalletta). Preliminary results reach estimates of efficiencies from each of the two (lattice-matched) sub-cells in excess of 21% per cell (predicted synergy of the two sub-cells in excess of 40%). Loss mechanisms at interfaces and quantum wells and their role in overall efficiency determination will also be included.

Advantages of the design are:

1. Solar spectrum matching in both visible and IR ranges through layer band gap-matching selection.
2. Lattice-matching and hence less carrier scattering (3) improved carrier transport due to GaAs (no mobility problems as in III-N-V hetero-solar cells). It is conceivable that even the 50% target of conversion efficiency will be reached with such a design.
3. No tunnel junctions (TJ) needed (1) the p-layer of the top cell is the same p-layer of the p-i-n superlattice, with no TJ need. Instead, the superlattice itself is a set of successive tunnel junctions (quantum well and barrier interfaces).

Heterostructure and (most recently) multijunction solar devices exhibit better performance in transport properties, when compared to bulk solar cells: especially in quantum well devices, photo-excitation causes carrier accumulation in discrete energy levels, with subsequent escape to the conduction band (minus recombination losses) via standard mechanisms such as tunneling, thermal escape or NNH conduction. Full spectrum absorption and triple junction solar cells have become key factors for high efficieny collection in PV structures of various geometries. Most recently, successful PV device designs have shown high efficiency values well above 30%, and efficiency levels in excess of 40% have been reached by means of triple junction metamorphic solar cells and under high sun concentration (good candidate for concentrated PV or CPV). Multijunction solar cells offer a great advantage over their bulk counterparts: by incorporating lattice-matched alloys, one may succeed in designing a device with more than one energy gaps thus increasing the number of absorbed solar photons. During the last decade, various groups have modeled and developed *multijunction* solar cells in order to increase overall collection efficiencies. Emphasis has been given in two types of PV devices (a) lattice-matched solar cells and (b) metamorphic (lattice-mismatched) solar cells. In particular, III–V multijunction solar cells have shown the greatest progress in overall efficiency.

5.8 The Concept

The broader impact of this project is a new design proposal for high efficiency solar cells. The target is to exceed 45% collection and thus to open the way to very efficient PV. It is more than clear that, once such a cell is realized, the field of concentration photovoltaics (CPV) will benefit greatly: solar cells with (a) record high efficiency values, (b) under several hundred suns (typically Fresnel optics at 500+ suns), and (c) small in size (low area hence less material) is already attracting interest for mass production in many places in the world. In recent years, it has been proposed by us a new design for a high efficiency and lattice-matched solar cell (HESC), where both visible and IR portions of the solar spectrum are absorbed according to the structure's geometric material arrangement: simultaneous absorption of both short and long wavelengths. In this on-going research enterprise the synergy, between a highly efficient triple junction cell and a highly efficient superlattice or a multi-quantum well region, is presented as a new and innovative way for further efficiency increase. It is well established by now, that triple junction solar cells are exceeding the upper threshold of collection efficiency to ever higher levels, namely above 38% with latest threshold at 40.8%. Currently, we are targeting a cell structure that will operate above the 40% threshold, with ultimate target the efficiency at or near 50%. Our proposal is based on a p-i-n bulk device model with three distinct areas, two of which are complete PV-heterostructures on their own, in other words these two regions could *stand alone* as two independent solar cell structures with quite acceptable performance (of the order of 21% and more as it has been demonstrated by our group recently). The power output of the PV composite device is a function of the individual power outputs from each sub-cell in the PV unit. A general formula of the *composite efficiency* will be discussed later on. Our group has already modeled a 21% triple junction AlAs/AlGaAs/GaAs solar cell. In parallel to that, our group has found more efficient options with various material combinations at higher yields. This task is in unison with other groups' work in the high efficiency solar cells. On the other hand, triple junction solar cells seem to lead the way to high efficiency PV especially in the area of CPV, where small cell area and therefore less material (hence lower material costs) may lead to high PV performance. The latter are triple junctions of lattice-matched and non-lattice matched III–V heterostructures with two tunnel junctions between the layers.

5.9 Current Research Objectives: A Proposed Guideline

1. Fully develop a theoretical model of PV composite PV devices by first principle calculations and computations based on realistic device parameters; propose a composite PV structure with two major cells: a triple junction and multi-layer tuned cell, with the prospect of high efficiency near 50%. Modeling tools include several established math software packages.
2. Seek for a composite PV device that combines properties of direct-gap crystalline semiconductors and absorption in the entire spectrum (as shown in Fig. 5.3 below), mainly in the visible and in the IR (NIR/IR) wavelength ranges, and which is configured as a two-part solar cell: a top triple junction and a multi-layer p-i-n bottom unit tailored to IR wavelengths. The solar spectrum (a 6,000 degree black body, shown in Fig. 5.4) offers the option of finding suitable band gaps for highest absorption. Material selection shows a blue shift in the absorption via wide gap materials as shown. On the other hand, low gap materials offer wavelength matching in the IR range.
3. Exploit the advantages of quantum wells grown on n-type or low-doped substrates. Seek for the synergy between a high efficiency triple junction of

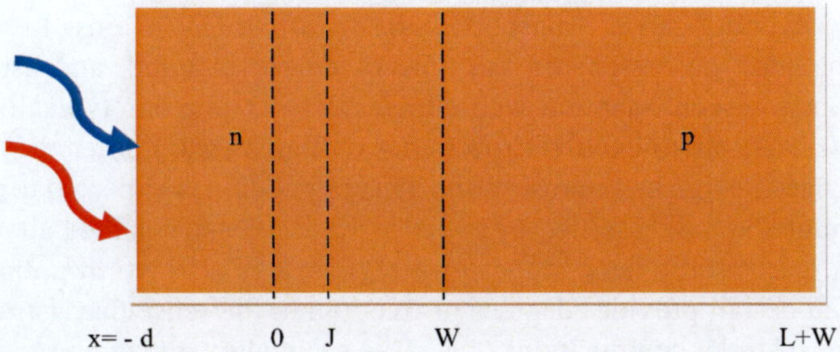

Fig. 5.3 Typical modeling geometry of a solar cell: w is the depletion width, J is the exact interface, L is the width of the p-region and d is the n-region (window layer), and w is the depletion width

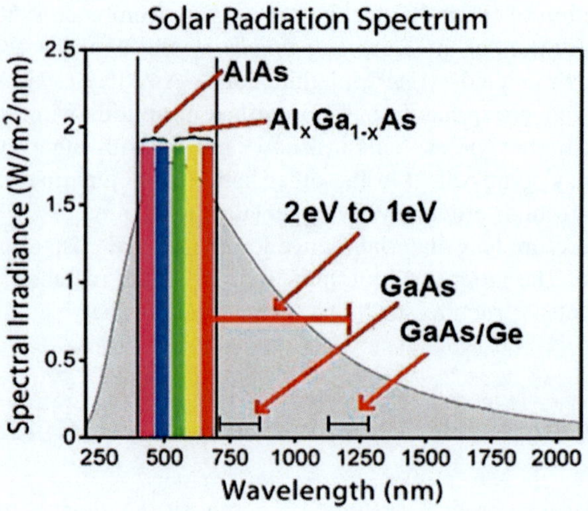

Fig. 5.4 Regions of the solar spectrum covered by the superlattice cell and the top cell (visible). The superlattice can be finely tuned at ~1 eV (via optical gap or energy distance between lowest electron and hole energy levels in quantum wells). Note the range for GaAs-Ge region of absorption

our own design and a MQW/superlattice structure (where quantum size effects are dominant). Superlattice structures in both cases mentioned above are at the designer's disposal, in the sense that appropriate quantum well geometries may lead to desired solar photons absorption.

4. Enhance cell performance by replacing the intrinsic region by a *finely-tuned lattice-matched MQW layer* via a wide gap – low gap sequence, but in such a way that *specific* range of wavelengths will have a chance to be absorbed. The latter region will be treated as a quantum trap of photo-generated electrons and holes. Photo-carriers trapped in quantum wells (medium with lower gap) may (a) recombine (b) escape from quantum wells into the conduction band of the wide gap medium, and/or (c) tunnel through thin barriers (if so desired). Thermionic emission, hopping conduction, and tunneling are dominant mechanisms of photo-carrier transport in heterostructures (against losses due to recombination processes). In the type of device in mind, and especially in its intrinsic region quantum wells, incident solar photons typically generate 10^{12}–10^{13} net photo-excited carriers per unit area (cm^2), after recombination effects have been taken into account. This population is expected to migrate to the conduction band assisted by the escape mechanisms named above and the built-in electrostatic field in the p-i-n region. It is also worth noting that the proposed design provides degrees of freedom in the sense that various lattice-matched material combinations (e.g., GaAs and alloys) may be considered for cell design, and at specific wavelength ranges.

5. Grow an **AlAs/AlGaAs/GaAs** lattice-matched n-n-p hetero-cell on top of a p-i-n quantum well based and lattice-matched cell. The **AlAs/AlGaAs/GaAs** option can reach collection efficiencies above 21%, if considered as a stand-alone solar cell. In addition, recent modeling of a triple junction **InP-GaAs-GaAs** cell, has shown 30% efficiency (1 cm^2 cell/one sun/1.5 AM conditions (Varonides, Spalletta 34th IEEE PVSC, June 7–12, 2009, Philadelphia). It is clear that an obvious advantage would be imminent if such a layer could be grown at the top of a p/i/n solar cell: the resulting proposed PV device includes two sub-cells in one. The proposed complete structures are: **n^+-AlAs/n-AlGaAs/p-GaAs/i/(GaAs/Ge)$_{MQW}$/n-(Ge)** and **n^+-InP/n-GaAs/p-GaAs) /i/(GaAs/Ge)$_{MQW}$/n-(Ge)**
6. Solve for thermionic currents in the intrinsic region; solar photons generate electron and hole pairs (EHP) when incident on a semiconductor layer. Photo-excited carriers may be trapped (confined) in quantum, wells and may either recombine or sustain a population in the eigen-energy levels. Existence of Ge-quantum wells serves this purpose: to confine a certain population of photo-excited electrons in the conduction band, out of which, a percentage is expected to escape to the conduction band of the surrounding material (GaAs). Given that electrons exhibit superior mobility values in un-doped GaAs, it is conceivable to expect improved transport properties in this GaAs-dominated PV device, over its III-N-V counterparts. Solving for thermal currents means (a) deriving an explicit expression of excess electrons $\delta n(x)$ per unit area (in any quantum well), (b) evaluating thermal currents out of each trap, (c) including recombination effects (mainly Auger and radiation recombination), and (d) solving the diffusion equation. The latter (diffusion of photo-electrons in the conduction band) involves a straightforward analytic method for solving the diffusion equation: by this method one is able to analytically predict carrier concentration in quantum wells (anywhere in the intrinsic region) and predict thermionic escape current density values, based on the excess carrier concentration $\delta n(x)$ (Varonides, Sze).
7. Solve for hopping conduction in the MQW region. NNH between quantum wells will be performed. NNH currents are non-zero in the intrinsic region. Carriers are expected to hop from energy level to neighboring energy levels (Ge-Ge sites). Work so far has shown that hopping currents (Varonides, Spalletta) are:

   (a) non-zero from site to site in the intrinsic region
   (b) strong functions of the Fermi level position relative to the band gaps in the intrinsic region
   (c) more specifically, we have concluded that hopping currents are related to Fermi level splitting at any two adjacent sites
   (d) depending on device geometry (quantum well width and barrier)
   (e) depending on temperature
   (f) conduction band discontinuity
   (g) density of states in the quantum wells.

8. Fine-tune at specific wavelengths in the IR region. Current research on crystalline solar cells focuses on a maximization of incident radiation absorption at all possible wavelength ranges, namely, from the visible to the near infrared (NIR). Currently, III-N-V's are being promoted for NIR absorption, but not without drawbacks: *low mobility* issues in (III-N-V) solar cells cause serious weakening of the advantages over NIR absorption. To overcome this problem we are proposing a superlattice structure in a standard p-i-n PV device, where the intrinsic region is replaced by a sequence of lattice-matched GaAs-Ge layers with pre-selected Ge (low gap) thickness chosen in such a way that overall optical gap values occur at or near 1 eV (Fig. 5.4). Quantum wells in the intrinsic region of the bottom sub-cell provide photo-excited carriers via the eigen-energy levels. Incident solar photons at long wavelengths can be collected (absorbed) by semiconducting layers with band gaps equal to the photons' energy. Quantum wells fit perfectly in the matter, causing increase of the band gap of the semiconductor (optical gap). Optical gap values can be tailored according to the incident photonic wavelength: eigen-energy levels in quantum wells are strong functions of well width (at any semiconductor interface there is a conduction band and a valence band discontinuity (ΔE_c and ΔE_v respectively)); the depth of the quantum well is equal to ΔE_c (for electrons). Thus, absorption can be finely tuned by selecting a priori the well width so that optical energy values are at desired wavelengths dictated by incident solar photons. So, first principle modeling and computations of Ge-quantum wells neighbored by GaAs layers (barriers) will be needed. As it has been realized in the wider literature, absorption at one electron-volt is desired (wavelengths near 1.24 μm). Absorption in the area of 1 eV is easily achievable by appropriate well width selection; this is the innovative idea: to (a priori) select optical gap energy values near 1 eV for IR absorption. An obvious advantage of the superlattice or MQW-based fine tuning is that different materials can be used (the more lattice-matched the better) for different optical gap-based absorption. Note also that eigen-energy levels in quantum wells (or quantum dots) may serve as second order absorbers (For instance, in a given QW arrangement, a second a hole-electron optical gap may correspond to 1.33 eV (very close to InP- band gap!), which translates to 0.932 μm absorption and so on (Fig. 5.5).
9. Select geometry of the quantum wells such that two (at the most) energy levels in the quantum wells are to be formed, namely, the ground state of electron-hole pairs at 1 eV, and the second state at the very *edge* of the GaAs layer conduction band this event has been shown to act in favor of *nearest neighboring hopping electrons*from site to site (QW). Thus a threefold advantage of the superlattice/MQW region is that (1) carriers are trapped and thermally escape to the conduction band, (2) NNH conduction becomes a second conduction mechanism, and (3) band gaps of other materials may be represented via energy levels in quantum wells.

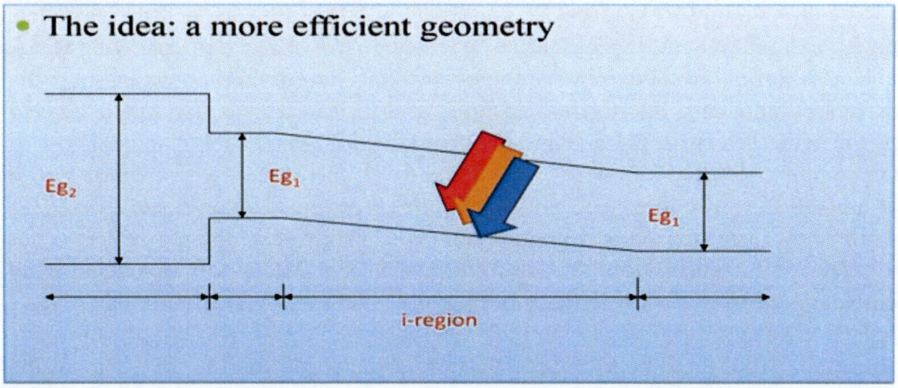

Fig. 5.5 A heterojunction p-i-n solar cell; high and low gaps serve for high and low energy or short and long wavelength photon absorption (visible and infrared respectively)

10. Calculate total efficiency by realizing the fact that two solar cells will eventually absorb input power according to their capability of photon absorption at the right portions of the solar spectrum. If P_{in} is the total input power, n_1, n_2 are the individual efficiency values of the two sub-cells involved in the device design, then the power output P_{o} is found from the following formula:

$$P_{\text{o}} = n_1 P_{\text{in}} + n_2 (P_{\text{in}} - P_{\text{o1}}) = (n_1 + n_2) P_{\text{in}} - n_1 n_2 P_{\text{in}}$$

Total efficiency for two sub-cell multijunction PV-device is:

$$n = n_1 + n_2 - n_1 n_2$$

(where) $\eta_1 = \frac{P_{\text{o1}}}{P_{in}}$.

In conclusion, the total current from the intrinsic region will be the sum of the thermionic and the NNH current components (minus recombination losses). Subsequent well width selection may lead to further refinement of solar photon absorption. Near IR and IR portions of the solar spectrum can be covered by suitable width selections, with equal amount of modeling effort (from the point of view of computations, it is a mere change of parameters for slightly different optical gaps). It is also interesting to note at this point that quantum well width could be modeled as a random variable, leading to a random distribution of optical gap values (as function of well width) and hence a smeared distribution of optical gap values and absorbed photon wavelengths, for the benefit of the PV device. The latter ensures IR photon absorption in the neighborhood of 1 eV and beyond (depending on the energy distance between eigen-energy levels in the quantum well), along with parallel minimization of electron capture, through NNH. In addition, the superiority of transport properties of the proposed quantum-PV device should be noted compared to its III-N-V "high" efficiency counterpart: our proposed MQW solar cell is mainly a GaAs device perturbed by thin Ge layers, and therefore this region exhibits much higher electron mobility. In the absence of tunneling (thick potential barriers) total currents are in essence the sum of thermionic and hopping current components, due to free electrons in the GaAs conduction band, assisted by the overall electrostatic field in the intrinsic region. The energy band diagram of the complete device is shown in Figs. 5.2 and 5.4: the existence of a window with "Anti-Reflecting-Coating" and surface texturing is included to minimize reflections, while shown are the top and bottom sub-cells. The goal in the top cell is either to retain visible absorption (AlAs/(Al) GaAs/GaAs at 21%) or to include a highly efficient triple junction cell (in this proposal, our own choice (InP/GaAs/GaAs at 30% efficiency). Wide-gap window materials (like AlAs, $Al_xGa_{1-x}As$) provide absorption from the blue to the red portion of the spectrum; the p-i-n part of the composite PV device covers the NIR and IR portion.

Figure 5.6 depicts a structure with various layers shown. Note also that in such a structure, there is no need for tunnel junctions for the carriers to migrate from region to region. Instead, excess photo-carriers from the top-cell will migrate to the p-type GaAs layer and they will drift in the intrinsic region assisted by the electrostatic field. Escaping photo-carriers from the intrinsic region will diffuse to the end of the device. Lack of tunnel junctions in the device is of fundamental importance over currently highly efficient multijunction solar cells. The mere fact that tunnel junctions are absent indicates:

1. less material to grow
2. less complexity in the structure overall
3. reduced scattering of drifting and diffusing carriers
4. reduced carrier trapping and recombination (carriers in MQW region separate from their corresponding holes as being away from the quantum wells)

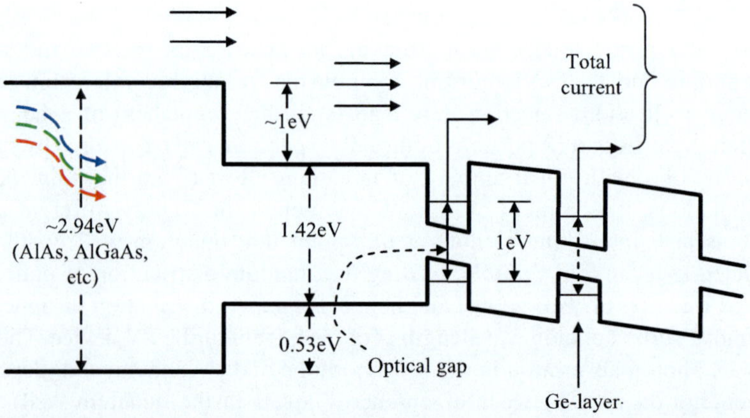

Fig. 5.6 Quantum PV structure: Bird's eye view of the device design proposed: shown are **(a)** different gap layers of the top cell, **(b)** GaAs-Ge layers in the intrinsic region, **(c)** optical gap at 1 eV, and **(d)** total conduction band free electrons due to thermionic escape and hopping conduction

5. faster growth conditions attainable
6. less fabrication costs.

Top cell design ensures visible wavelength absorption (via AlAs-AlGaAs-GaAs double junction cell) and the bottom cell provides absorption in the IR or near 1 eV (depending on the eigen-energy tuning in via the quantum wells).

5.10 To Probe Further

Cell structures of the type depicted above are promising state of the art PV devices with high currents (in excess of $30\,\mathrm{mA/cm^2}$ and increased open-circuit voltage (OC) values: near one volt per subcell under one sun). The latter notion related to the fact that the two sub-cells are considered to be connected in series, where the OC voltages are in essence in excess of two volts. This would lead to cumulative collection efficiency values well in excess over 40%; for instance: for a very realistic fill factor (FF) near 80%, for final matched currents in the neighborhood of $25\,\mathrm{mA/cm^2}$ *(a conservative value for such high-current structures)*, and for OC voltage near 2 Volts, the overall one-sun efficiency of the cell (as the one shown by Fig. 5.7) would exceed the 40% current threshold! Optimization of matching currents between the sub-cells needs to be explored further; superlattice based cells

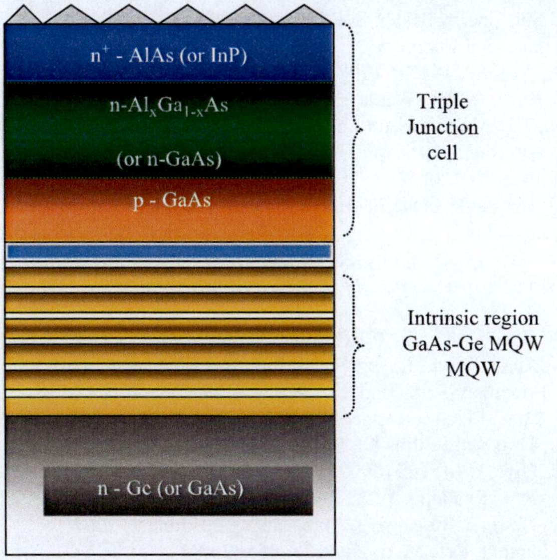

Fig. 5.7 Proposed PV-structure with two sub-cells: bulk top lattice-matched structure for the visible, and bottom cell structure for NIR, IR photons. The region between the two cells could be a tunnel junction to ensure current matching between the two cells

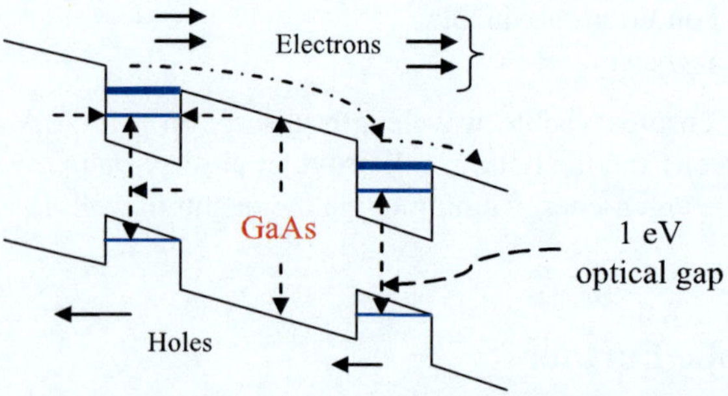

Fig. 5.8 Fine tuning at specific wavelengths can be achieved through quantum size effects in the quantum traps of the intrinsic region. Electrons' thermionic escape and NNH conduction are shown at the conduction band. Note the eigen-energy levels in Q W's: by selecting the Ge-layer width, second order levels can reach the conduction band edge of GaAs

are known to carry high current and coexistence of such layers in a series connected fashion, may lead to substantial voltage increase and hence to higher than 40% efficiencies (Fig. 5.8).

References

1. M. Yamaguchi, Sol. Energy Mater. Sol. Cells **90**, 3068–3077 (2006)
2. M. Yamaguchi et al., Sol Energy **82**, 173 (2008)
3. R.R. King et al., APL **90**, 183516 (2007)
4. J.F. Geisz, S. Kurz, M.W. Wanlass, J.S. Ward, A. Duda, D.J. Friedman, J.M. Olson, W.E. McMahon, T.E. Moriarty, and J.T. Kiehl, Appl. Phys. Lett. **91**, 023502 (2007)
5. T. Kirchartz, U.w.e. Rau, et al., Appl. Phys Lett. **92**, 123502 (2008)
6. T. Mei, J. Appl. Phys. **102**, 053708 (2007)
7. G.F.X. Strobl, T. Bergunde, et al., in *4th World Conference on Photovoltaic Energy Conversion*, Hawaii, 8–12 May 2006
8. A.C. Varonides, R.A. Spalletta, Phys. Stat. Sol. **5**, (2) 441 (2008)
9. G.F.X. Strobl et al., in *Proceedings of the 7th European Space Power Conference*, 9–Italy, 13 May 2005
10. H.L. Cotal, R.R. King et al., 28th IEEE Photovoltaic Specialists Conference, p. 1316 (2000)
11. T. Kieliba, S.W. Riepe Warta, J. Appl. Phys. **100**, 093708 (2006)
12. J.F. Geisz, D.J. Friedman, Semicond. Sci. Technol. **17**, 789 (2002)
13. W. Hant, IEEE Trans. Electron Devices **26**, (10) 1573 (1979)
14. A.C. Varonides, Thin Solid Films **89**, 511–512 (2006)
15. A.C. Varonides, Phys. E **14**, 142 (2002)
16. A.C. Varonides, R.A. Spalletta, W.A. Berger, in *European Materials Society, Spring Meeting 2005, Palais des Congrès,*Strasbourg, France, May 28–June 1, 2007
17. E. Istrate, E.H. Sargent, Rev. Mod. Phys. **78**, 455 (2006)
18. H.E. Runda et al., Nanoscale Res. Lett. **1**, 99 (2006)
19. L. Lazzarini et al., Micron **31**, 217 (2000)
20. M. Razeghi, J. AP **23**, 141 (2003)
21. H.W. Li, B.E. Kardynal, et al., Appl. Phys. Lett. **93**, 153503 (2008)

22. S.M. Sze, *Physics of Semiconductors* (Wiley and Sons, NY, 1981)
23. A.C. Varonides, Thin Solid Films **511–512**, 89–92 (2006)
24. A.C. Varonides, in *Proceedings of the 3rd Workshop on Nano-sciences & Nanotechnologies*, (Aristotle University of Thessaloniki, Greece, 2006), p. 46
25. A.C. Varonides, in *Thin Film and Nano-structured Materials for Photovoltaics, Reprinted from Thin Films*, ed. by A. Slaui, A. Jager Waldau, J. Poortmans, V. Brabec, E-MRS 2006, EMRS Proceedings 180, (Elsevier), Nice, France, p. 89
26. A.C. Varonides, in Hopping conduction III-N-V hetero-junctions, *Presented at the European Materials Society*, Strasbourg, France, May 31 2007
27. A.C. Varonides, R.A. Spalletta, W.A. Berger, in *European Materials Society, Spring Meeting 2005, Palais des Congrès,*Strasbourg, France, May 28–June 1 2007
28. A.C. Varonides, R.A. Spalletta, W.A. Berger, in *E-MRS, European Materials Society, Spring Meeting 2005, Palais des Congrès,*Strasbourg, France, May 28–June 1 2007
29. A.C. Varonides, R.A. Spalletta, Phys. Status Solidi (c) **5**, (2) 441–444 (2008)
30. A.C. Varonides, R.A. Spalletta, W.A. Berger, in *Proceedings of the World Renewable Energy Congress (WREC) X and Exhibition*, 19–25 July 2008, Scottish Exhibition & Conference Centre, Glasgow, Scotland, UK

Chapter 6
Thin Film Deposition and Nanoscale Characterisation Techniques

Spyridon Kassavetis, Christoforos Gravalidis, and Stergios Logothetidis

Abstract In this chapter the basic categories of the thin film deposition techniques are presented. The *rf* magnetron sputtering (MS) deposition conditions and their effect to the optical, nanostructural and nanomechanical properties of: (a) single layer and multilayer, hard and soft carbon-based thin films grown on rigid Si substrate and (b) AlO_x thin films grown on flexible polyethylene terephthalate (PET) substrate are presented and discussed. Finally the Spin coating technique for the development of thin films from the liquid phase and the solvent effect to the PEDOT:PSS thin film thickness are given.

6.1 Introduction

A layer of a material that covers the surface of another material and has a thickness from few nanometers to a micrometer is called thin film. Materials in the form of thin films are very important for the contemporary and future research and technology and have lead to the miniaturization of existing devices and to the development of novel ones. The thin films cover a very wide range of applications from relatively simple ones, like decoration in architecture and jewelry, protection of materials surfaces from unfriendly/corrosive environments to more sophisticated in optoelectronics, MEMS, NEMS, and micro/nanoelectronics.

Simultaneously with the invasion of the thin films to the technology of modern devices, the thin film deposition techniques are also evolving rapidly in the last decades, in order to meet the market needs for high quality multifunctional materials that combine exceptional mechanical, optical, electrical, and magnetic properties.

In addition the characterization techniques, real-time, in-situ, and ex-situ have been developed in order to provide valuable information about the thin film growth

K. Spyridon (✉) · G. Christoforos · L. Stergios
Physics Department, Lab of "'Thin Films - Nanosystems & Nanometrology (LTFN)",
Aristotle University of Thessaloniki, Thessaloniki, Makedonia, Hellas
e-mail: skasa@physics.auth.gr; cgrava@physics.auth.gr; logot@auth.gr

S. Logothetidis (ed.), *Nanostructured Materials and Their Applications*,
NanoScience and Technology, DOI 10.1007/978-3-642-22227-6_6,

mechanisms, the nanostructure, the properties, the interface quality, the adhesion of the thin film to the substrate and to control the deposition process.

In the following paragraphs thin film deposition techniques from vapor and liquid phase are used for the growth of hard and soft amorphous carbon thin films and organic thin films, respectively. Characterization techniques, sensitive in nanoscale phenomena and structure are used to study the effect of the deposition conditions to the optical and nanomechanical properties of the thin films.

6.2 Methods and Results

6.2.1 Thin Film Deposition Techniques

In general, the thin film deposition techniques are divided in three major categories: the Physical Vapor Deposition (PVD) [1, 2], the Chemical Vapor Deposition (CVD) [3], and the Wet deposition techniques. In Fig. 6.1 the thin film deposition techniques as well as their subcategories are presented.

In the following paragraphs we focus on thin films deposited by using Physical Vapor Deposition, and Wet deposition techniques. In the case of the PVD the discussion centers at the magnetron sputtering of carbon-based thin films, such as amorphous carbon and hydrogenated amorphous carbon.

6.2.2 Physical Vapor Deposition: Magnetron Sputtering

The deposition of a thin film by using the magnetron sputtering (MS) technique is made in two steps. In the first step atoms are ejected (sputtered) from the surface of a solid (the target material) and in the second one the sputtered particles are deposited on a substrate [2]. In Fig. 6.2 the most usual geometry of a sputtering system is

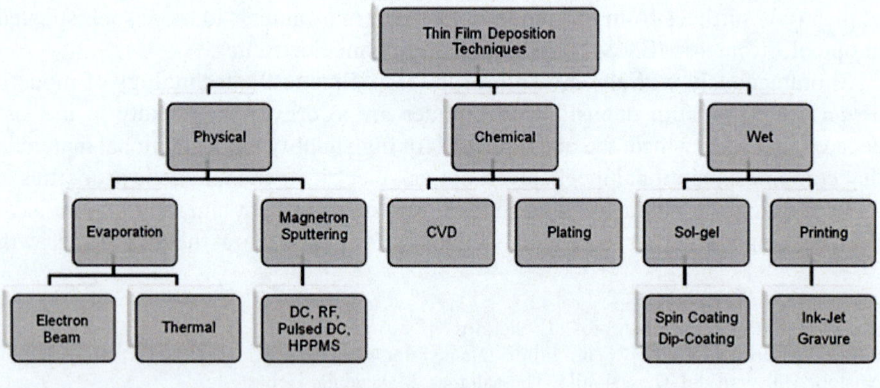

Fig. 6.1 Overview of thin film deposition techniques

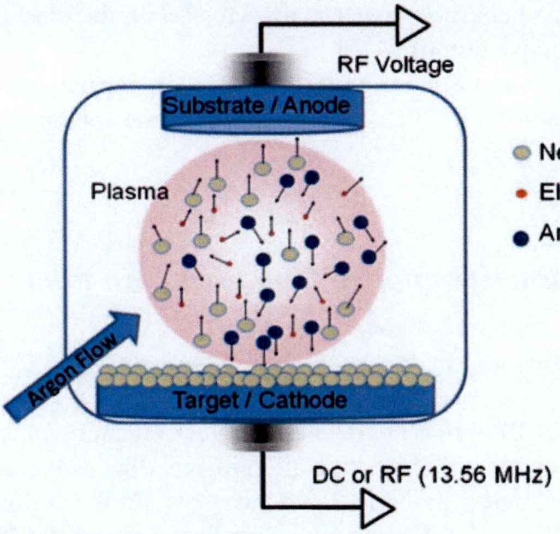

Fig. 6.2 Schematic representation of the magnetron sputtering process

presented. In the cathode the ultra pure target material is placed (usually 99.999% pure), while the substrate is placed in the anode.

The sputtering process is based on phenomena occurred when an energetic gas ion impacts the surface of a solid (target material). The ion-surface interactions can lead to successive sputter of surface atoms or to un-successive events such as scattering or absorption and implantation of the ion. In a successive event the incident ions set off collision cascades in the bombarded solid and the surface atom gets sputtered when the energy of these cascades at the solid surface is above the surface binding energy. Thus, the sputtering efficiency is defined as the average number of atoms ejected from the target per incident ion and is given by the sputtering yield S:

$$S = \frac{\text{Number of sputter atoms}}{\text{Incident ion}} \tag{6.1}$$

The sputter yield depends on the ion energy, incident angle, the ion and target atoms masses, and the surface binding energy of target material atoms. In the literature typical values of S are between 10^{-5} and 10^{3} [4].

In MS, magnetrons, which are placed beneath the high purity target material, are used to induce strong electromagnetic field, which traps the electrons close to the target surface. These electrons enable plasma sustainability and ionization of more neutral gas atoms near the target surface. The latter contributes to the sputtering yield increase and as a consequence to higher deposition rate. The MS divisions based on the power supply are the following:

- *dc* MS, which is used for the deposition of thin films from conductive materials/targets (metals) [1, 2].
- *rf* MS, applied for the deposition of conductive, semiconductive, and insulating thin films for corresponding materials/target [1, 2].

- Pulsed *dc* MS, in which a pulsed *dc* power is used to sustain the plasma during the growth of non-conductive thin films [5].
- High Power Pulsed Magnetron Sputtering (HPPMS), high plasma densities are obtained via much higher total ion flux and ionization of the sputtered material close to 70% [6, 7].

6.2.3 Nanoscale Characterization of Sputtered Thin Films

6.2.3.1 Spectroscopic Ellipsometry

Spectroscopic ellipsometry (SE) is an accurate, non-destructive, nanoscale sensitive technique that is used for the optical characterization of metallic, semiconductive, dielectric inorganic, organic, and polymeric thin films [8, 9]. In the following the SE measurements in the NIR-Vis-fUV spectral region are analyzed to derive the optical properties, the electronic transitions, and the bonding of carbon-based thin films.

Generally, in the case of a thin film grown on a substrate, a three-phase model is considered: the semi-infinite ambient [medium (0)], the thin film [medium (1)] with thickness d and the substrate [medium (2)]. The complex reflectance ratio is defined as:

$$\langle \rho \rangle = R_p / R_s, \tag{6.2}$$

$$\tilde{R}_p = \frac{\tilde{r}_{01p} + \tilde{r}_{12p} e^{i2\beta}}{1 + \tilde{r}_{01p}\tilde{r}_{12p} e^{i2\beta}} \tag{6.3}$$

$$\tilde{R}_s = \frac{\tilde{r}_{01s} + \tilde{r}_{12s} e^{i2\beta}}{1 + \tilde{r}_{01s}\tilde{r}_{12s} e^{i2\beta}}, \tag{6.4}$$

where r_{01i} and r_{12i} are the *Fresnel* reflection coefficients for the interfaces between medium (0) and (1), and medium (1) and (2), respectively, and the phase angle β is given by:

$$\beta = 2\pi \left(\frac{d}{\lambda}\right) \sqrt{n_1^2 - n_0^2 \sin^2 \theta}, \tag{6.5}$$

where λ is the wavelength, θ is the angle of incidence, and n_0 and n_1 are the complex refraction indices of the ambient and the film, respectively. The phase angle β diminishes in the energy region of high absorption, leading to $R_i = r_{01i} = r_i$, ($i =$ p, s) and consequently to $\langle \rho \rangle = \rho$. Therefore, at the absorption bands only the information of the thin film optical properties are obtained.

The real $\varepsilon_1(\omega)$ and the imaginary $\varepsilon_2(\omega)$ parts of the dielectric function are connected via the Kramers–Kronig relations (are based on the causality principle [11]):

$$\varepsilon_1(\omega) = 1 + \frac{2}{\pi} P \int_0^{\infty} \frac{\omega' \varepsilon_2(\omega')}{\omega'^2 - \omega^2} d\omega' \tag{6.6}$$

$$\varepsilon_2(\omega) = -\frac{2\omega}{\pi} P \int_0^{\infty} \frac{\varepsilon_1(\omega') - 1}{\omega'^2 - \omega^2} d\omega', \tag{6.7}$$

where P is the principal value of the integral close to the material's electronic resonance ($\omega' = \omega$) and

$$\varepsilon_1(\omega = 0) = 1 + \frac{2}{\pi} P \int_0^{\infty} \frac{\varepsilon_2(\omega')}{\omega'} d\omega' \tag{6.8}$$

is the static dielectric function – the material strength (deviation from the strength of vacuum $\varepsilon_0 = 1$), that describes all the losses through-out the electromagnetic spectrum, due to the absorption in the material.

In the case of materials thinner than the light penetration depth, such as the thin films grown on a bulk substrate, SE provides the pseudo-dielectric function $\langle \varepsilon(\omega) \rangle$, which contains contribution from both the substrate (optical properties) and the thin film (optical properties and thickness). Analysis of the measured pseudo-dielectric function $\langle \varepsilon(\omega) \rangle = \langle \varepsilon_1(\omega) \rangle + i \langle \varepsilon_2(\omega) \rangle$ in the NIR-Vis-fUV energy region can lead to the calculation of the thin film thickness, with sub-nanometer resolution, and enables the study of the thin film electronic band structure and the electronic transitions, since the $\varepsilon_2(\omega)$ is directly related to the joined density of states for interband absorption [9, 11].

A Phase Modulated Spectroscopic Ellipsometer (Horiba Jobin-Yvon) was used to acquire the SE measurements in the NIR-Vis-fUV spectral region (0.7–6.5 eV) at an angle of incidence of 70°, with a step of 20 meV [12].

For the analysis of the $\langle \varepsilon(\omega) \rangle$, a two-layer material that consists of the a-C:H layer with thickness d on top of the c-Si substrate was considered. A Tauc–Lorenz (TL) oscillator, which is widely used to study the optical properties of the amorphous semiconductors, was used to describe the optical response of the a-C:H film [13, 14]. The TL oscillator is a combination of the classical Lorentz dispersion relation and the Tauc joined density of states in the proximity of the fundamental optical gap ω_g [9, 11]. The analytical relations that describe the TL oscillator are the following:

$$\varepsilon_2(\omega) = \begin{cases} \dfrac{A\omega_0 C(\omega - \omega_g)^2}{(\omega^2 - \omega_0^2)^2 + C^2\omega^2} \dfrac{1}{\omega} & \omega > \omega_g \\ 0 & \omega \le \omega_g \end{cases} \tag{6.9}$$

$$\varepsilon_1(\omega) = \varepsilon_\infty + \frac{2}{\pi} \mathrm{P} \int_{\omega_g}^{\infty} \frac{\xi \varepsilon_2(\xi)}{\xi^2 - \omega^2} d\xi. \tag{6.10}$$

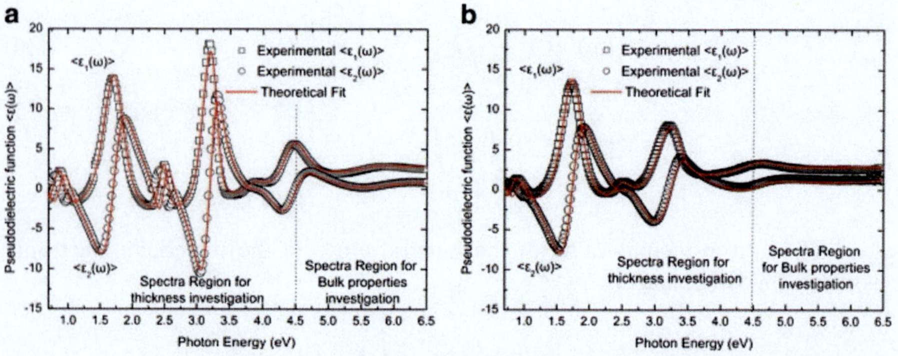

Fig. 6.3 Pseudo-dielectric function $\langle\varepsilon(\omega)\rangle$ (points), of two representative a-C:H thin films deposited onto c-Si substrates under (**a**) floating bias voltage (+11 V) and (**b**) bias voltage of −100 V. The solid lines represent the theoretical fit of the measured spectra (points)

The imaginary part $\varepsilon_2(\omega)$ of the dielectric function is the result of the multiplication of the Lorentz oscillator and the Tauc joint density of states equations, while the real part $\varepsilon_1(\omega)$ is calculated by the imaginary part $\varepsilon_2(\omega)$ via the Kramers–Kronig integration [9, 11]. In (6.1) the ω_g is the fundamental band gap energy, A is the amplitude, ω_0 is the energy position of optical transition (namely the Penn gap), and C the broadening term of the optical transition [13,14]. The ε_∞ value is above unity due to the electronic transitions contribution occurred at higher photon energies, not covered by the measured spectra. A nonlinear Levemberg–Marquandt minimization algorithm was used for the regression process and the calculation of the best-fit parameter.

In Fig. 6.3 a and b, the measured spectra in the nIR-Vis-UV spectra region together with the fitting results (solid lines) are presented. The agreement between the experimental and the theoretically calculated $\langle\varepsilon(\omega)\rangle$ proves the validity of the performed analysis. Two TL oscillators (2-TL model) were used to describe and distinguish the contributions of the π–π* and σ–σ* interband transitions. The thickness of the a-C:H films of set #A was estimated around 260 nm.

In Fig. 6.4, the results of the above analysis are presented. The energy band gap ω_g, and energy of π–π* and σ–σ* interband electronic transitions are plotted versus the applied negative voltage (V_b) to the substrate. The ω_g values were found to decrease with increasing the V_b. For $V_b < 20$ V, the a-C:H films are characterized by higher optical transparency ($\omega_g \approx 2$ eV) compared to those grown for $V_b > 20$ V($\omega_g \approx 1.5$ eV for $V_b = -100$ V). The observed change to the ω_g can be attributed to the more intense ion bombardment of the growing a-C:H films triggered by the higher values of applied negative V_b. The bombardment with ions, neutral atoms, and species during the growth of the a-C:H film surface, promotes the formation of sp^3-type bonding structures among carbon atoms. Simultaneously, the sp^2-bonded carbon is formed and contributes to the relaxation of the intrinsic compressive stresses occurred due to the expansion of sp^3-carbon

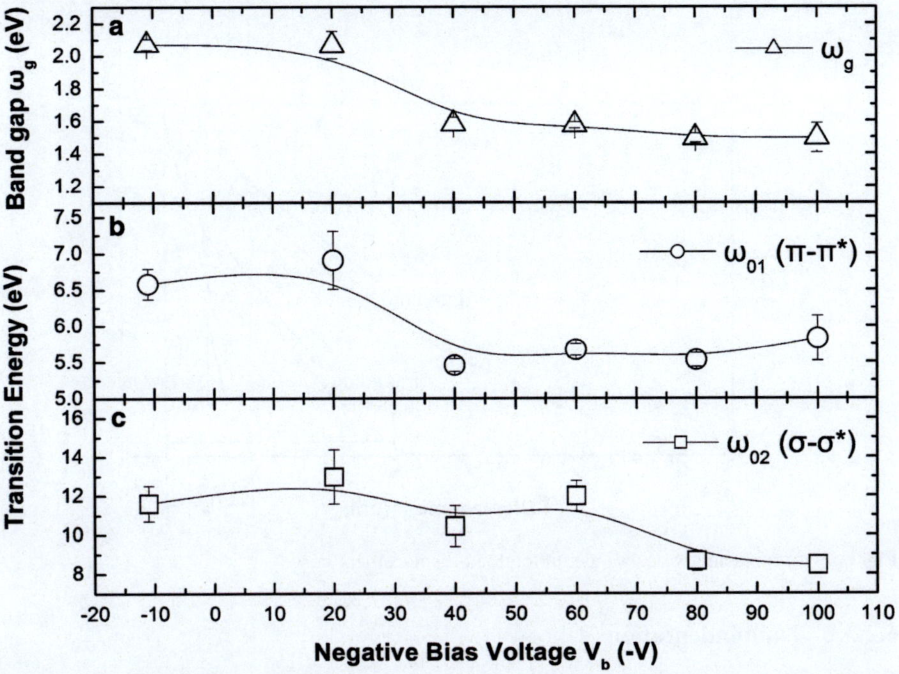

Fig. 6.4 (**a**) The fundamental gap ω_g (**b**) the π–π* and (**c**) the σ–σ* interband electronic transitions as a function of the negative bias voltage V_b applied to the substrate during the deposition of the a-C:H films

network [15–19]. According to these models, the ω_g reduction (the decrease of the optical transparency) is explained by an increase in both the size and the degree of distortion of sp^2 clusters (aromatic rings and olefinic chains) that are embedded in the sp^3-type disordered a-C:H network (graphitization of the a-C:H films).

The same behavior has been also reported for a-C:H thin films grown by *rf* MS, with almost similar experimental conditions[20], dual electron cyclotron resonance, [15, 18] and filtered cathodic vacuum arc [21].

The π–π* interband transition energy (ω_{01}) was found to decrease from 6.6 eV to 5.8 eV while the V_b increases from −11 V (floating) to 100 V. The latter ω_{01} value tends to that of the crystalline graphite ($\omega_{01} \approx 4.5$ eV). It should be also noted that the π–π* interband transition energy ω_{01} of the non-hydrogenated a-C thin films varies between 4 and 4.5 eV [20, 22]. This ω_{01} reduction can be attributed to the progressive reduction of the π–π* transition energy between the bonding and anti-bonding splitting of the π states [15]. Moreover, the analysis showed that the calculated σ–σ* interband transition energy ω_{02} reduces to $\approx$ 10 eV for the a-C:H films deposited with higher values of V_b (Fig. 6.6b). This is correlated to the favorable formation of the sp^3 bonded carbon structures triggered by the more intense ion bombardment [15, 19, 23].

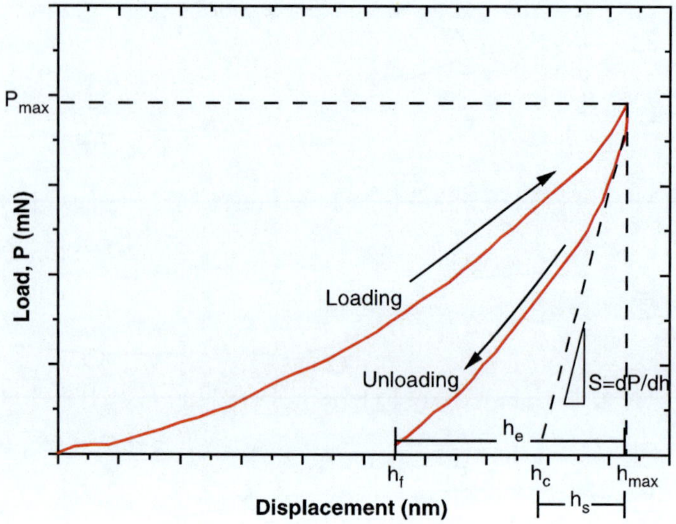

Fig. 6.5 A representative nanoindentation load-displacement curve

6.2.3.2 Nanoindentation

Nanoindentation (NI) is the evolution of the Brinnel hardness tests [24]. The prefix "nano" is justified by both the vertical displacement and the normal load applied by the indenter, as well by the indenter geometrical characteristics. So, in NI the indenter is usually a very sharp, triangular pyramid (Berkovich type) diamond with less than 50 nm tip roundness. The indenter vertically penetrates into the material, by controlling either the normal applied load with less than microNewton resolution or the displacement with nanometer resolution [25, 26].

The result of a NI measurement is the Load-Displacement (P-h) curve, which contains the loading and the unloading parts. The loading can be accomplished either by controlling the indenter displacement or the normal applied load. In Fig. 6.5 a typical NI P-h curve is presented. The NI P-h curve can be analyzed in order to estimate the elastic modulus (E) and the hardness (H) of a thin film or a bulk material, as well as to study the adhesion of a thin film on a substrate, the deformation due to creep, the phase transformations induced due to the applied pressure by the indenter.

The basic equations that give the E & H of a material are listed below:

$$\frac{1}{E_r} = \frac{1 - v_{\mathrm{sa}}^2}{E_{\mathrm{sa}}} + \frac{1 - v_i^2}{E_i}, \tag{6.11}$$

$$H = \frac{P_{\max}}{A}, \tag{6.12}$$

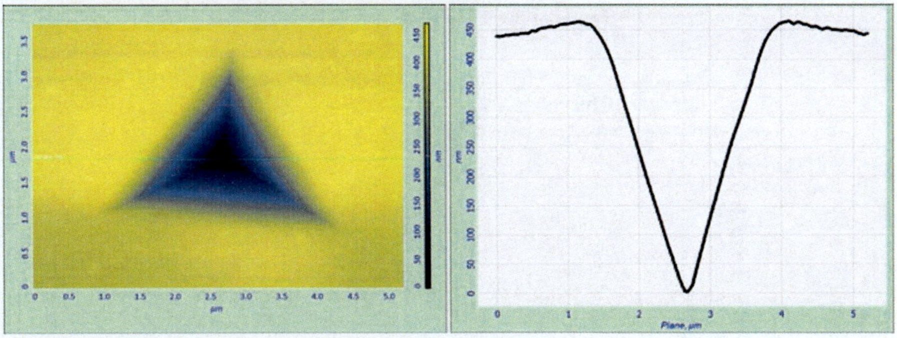

Fig. 6.6 The Berkovich indenter imprint on aluminum surface (**a**) and its profile at $h_{\max}$ (**b**)

where the subscripts *sa, i* correspond to the sample and the indenter, respectively, *r* is the reduced modulus, *v* is the Poisson ratio, $P_{\max}$ the maximum applied load, and A the contact area. The contact area in the case of a triangular pyramid (Berkovich type) indenter is given by the following equation:

$$A(h_c) = 24.5 \cdot h_c^2 + C_1 h_c + C_2 h_c^{1/2} + C_3 h_c^{1/4} + \ldots . + C_8 h_c^{1/128}, \qquad (6.13)$$

where h_c is the contact depth and the parameters C_i are estimated using the procedure presented in the Oliver–Pharr (O–P) paper [25]. In Fig. 6.6, the imprint of a Berkovich type indenter on Aluminum and its profile at the $h_{\max}$ are presented.

The O–P model is the most popular for the analysis of the NI P-h curve. According to the O–P model for the elastic contact, the initial portion of the unloading of an elastic medium after the elastic contact with a rigid indenter (flat, sphere, cone, paravoloid of revolution) follows the exponential rule:

$$P = c(h - h_f)^n, \qquad (6.14)$$

where P is the load, h is the displacement, c and n are constants related to the material and to the indenter geometry, respectively.

In Fig. 6.7 $h_{\max}$ is the maximum displacement of the indenter resulted at the maximum load ($P_{\max}$), h_f is residual contact depth after removal of the applied load (unloading), h_c is the contact depth between the rigid indenter and the sample, h_e is the final elastic recovery of the sample's surface, and h_s is the elastic displacement of the surface, which is not in contact with the diamond indenter.

The contact depth h_c is given by (6.15) as we can also see in Fig. 6.7

$$h_c = h_{\max} - h_s \qquad (6.15)$$

The h_s is estimated using the Sneddon analysis [27] and the (16) gives the contact depth h_c:

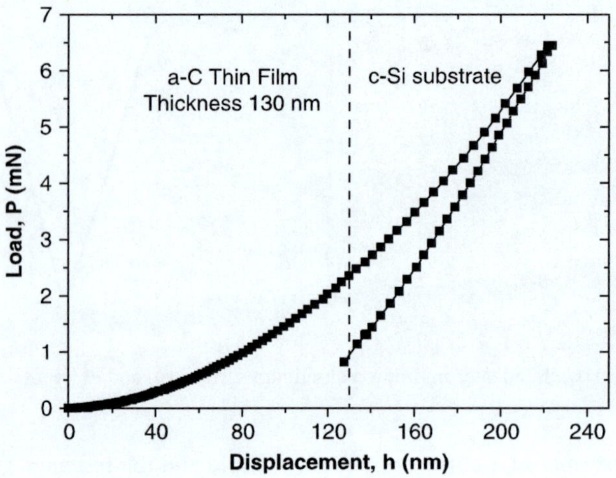

Fig. 6.7 Representative nanoindentation load-displacement curve from the single layer a-C/Si (001) sample

$$\left.\begin{aligned} h_s &= \frac{\pi - 2}{\pi}\left(h_{\max} - h_f\right) \\ \varepsilon &= 2\frac{\pi - 2}{\pi} \\ h_{\max} - h_f &= \frac{2P}{S} \end{aligned}\right\} \Rightarrow h_c = h_{\max} - \varepsilon\frac{P_{\max}}{S} \tag{6.16}$$

where ε is a coefficient, whose value depends on the indenter geometry (1 for a flat punch indenter, 0.72 for conical, and 0.75 for spherical). The Stiffness (S) is given by the slope of the elastic unloading in the P-h curve:

$$S = \frac{dP}{dh}. \tag{6.17}$$

The stiffness is directly related to the reduced elastic modulus via (18):

$$S = \left(\frac{2}{\sqrt{\pi}}\right) \cdot \sqrt{\mathrm{A}} \cdot E_r. \tag{6.18}$$

As a consequence the analysis of the P-h curve using the O–P leads to the estimation of the H & E of a material.

In dynamic nanoindentation (DNI) the contact stiffness is continuously measured versus the vertical displacement of the indenter. Thus, the E and H are also estimated continuously from the material surface till the maximum penetration depth [25, 28–30]. DNI is very useful especially in the case of the thin films, because it can provide valuable information about the variation of the mechanical properties of the inhomogeneous thin film/substrate material system. The DNI measurement is made at the loading portion of the P-h curve when a small AC driven force is superimposed to the applied force. This causes oscillated displacement of the indenter in the same frequency with that of the force but with difference in the phase (ϕ). The indenter oscillated force and displacement are very small (few μN and 2–4 nm, respectively) and do not influence the deformation of the sample. The ϕ between the force and the displacement is measured by a high precision phase sensitive lock-in amplifier.

The differential equation of the forced oscillation is the following:

$$F_o \sin \omega t - D\frac{dh}{dt} - K_s = m\frac{dh^2}{dt^2}, \tag{6.19}$$

where F_0 is the amplitude and ω the frequency of the oscillating force, dh is the instant displacement, D is the damping coefficient, K_s is the stiffness of the indenter supporting system, m is the mass of the indenter and S is the contact stiffness.

The contact stiffness can be calculated using the following equations:

$$\left|\frac{F_0}{h_0}\right| = \sqrt{(S + K_s - m\omega^2)^2 + \omega^2 D} \tag{6.20}$$

$$\tan \varphi = \frac{\omega D}{S + K_s - m\omega^2}. \tag{6.21}$$

From the contact stiffness S, both the elastic modulus and the hardness can be calculated using (11, 12, and 21).

6.2.3.3 Nanomechanical Properties of Thin Films

Thin Films Harder than Their Substrate

NI is nowadays widely used to measure the hardness and the elastic modulus as well as to study the deformation modes of nanostructured thin films. Especially the DNI is used to measure H & E versus the contact penetration depth.

In Fig. 6.7 the DNI P-h curve is presented. The diamond Berkovich indenter penetrated 219 nm into the a-C thin film (130 nm thick) by applying 6.28 mN normal applied load. The indenter passed through the thin film to the substrate and the contact is elastic/plastic. Both the loading and the unloading part of the curve are continuous. Thus neither delamination nor cracking of the thin film was occurred.

The hardness of the a-C thin film was continuously measured with contact depth. The H values, which were calculated after analysis of the P-h curve loading portion, are presented in Fig. 6.8. These are the average values calculated from ten nanoindents made in different areas of the sample and the error is the standard deviation. The H values increase as the Berkovich diamond indenter penetrates deeper into the sample and the contact goes from elastic to elastic/plastic. The H values decrease for h $\geq$ 20 nm due to so-called substrate effect [30–32].

In order to subtract the substrate effect,we assume that the mechanical properties of the c-Si substrate affects the measured hardness from h = 0. This is also

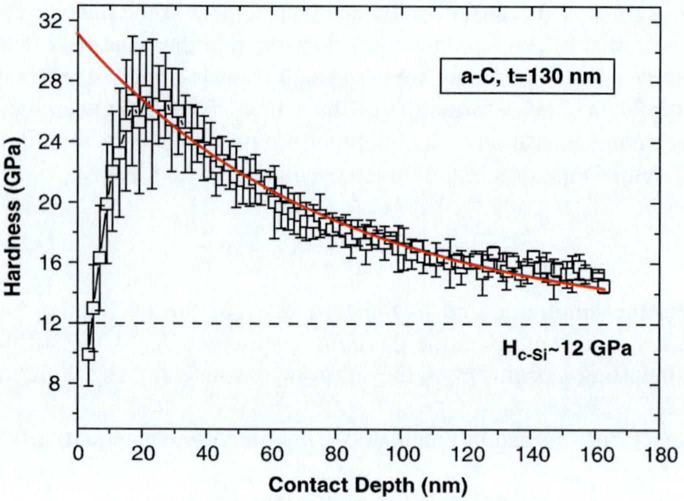

Fig. 6.8 The average hardness values calculated from ten nanoindents to the a-C thin film

supported from the absence of any clear value plateau. In order to estimate the thin film hardness at the surface (h = 0) the following equation from the Bhattacharya–Nix model is used [32]:

$$H = H_s + \left(H_f - H_s\right) \exp\left[-\frac{h_c}{d}\frac{H_f/H_s}{\alpha/E_s^{1/2}}\right], \tag{6.22}$$

where the subscripts f and s correspond to the thin film and to the substrate hardness and elastic modulus (E_{c-Si} = 160–180 GPa and H_{c-Si} = 12 GPa), respectively, and α is a coefficient related to the difference between the hardness values of the thin film and the substrate.

The solid line in Fig. 6.8 comes from the fitting to hardness experimental data using (6.22). So, the hardness to the thin film surface is 32 GPa. In the same way the real hardness of several a-C thin films was estimated. These thin films were either single layer rich in sp^3 and sp^2 hybridized carbon bonds or a-C multilayers consisting of alternating sp^3 and sp^2 hybridized carbon bonds layers. The a-C multilayers were deposited using *rf* MS. The rich in sp^3 hybridized carbon bonds was a-C layers were grown with applied negative bias voltage at the substrate, while the rich in sp^2 hybridized carbon bonds were grown without (floating) negative bias voltage at the substrate.

In Fig. 6.9 the a-C thin films hardness is plotted versus the percentage of the sp^3 hybridized carbon bonds [8]. A linear increase of the hardness with the increase of the sp^3 hybridized carbon is observed [31].

In the case of the a-C multilayers, the H increase follows the relation H_f = $0.87 \times$ (% sp3 hybridized carbon bonds content) GPa. Also the multilayer thin films appear to be much harder compared to single layer 30 nm and 130 nm thick thin

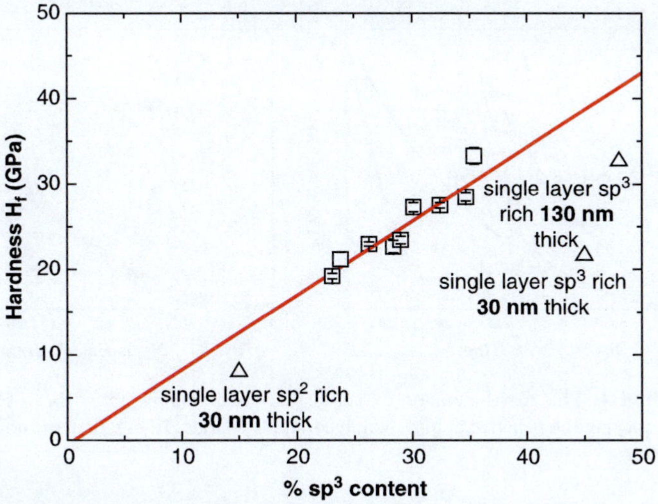

Fig. 6.9 The hardness of single layer and multilayer thin films versus their content in sp^3 hybridized carbon bonds

films, rich in sp^3 hybridized carbon bonds ($H_f = 24$ GPa, % sp^3 content ≈ 45 and $H_f = 32$ GPa, % sp^3 content ≈ 47) despite of their lower sp^3 hybridized carbon bonds content.

Thin Films Softer than Their Substrate

In this section the case of thin films softer that the substrate is discussed. Also the DNI results will be compared with those coming from atomic force acoustic microscopy (AFAM). The thin films are hydrogenated amorphous carbon (a-C:H) deposited onto c-Si using *rf* MS in high vacuum (P$<10^{-7}$ mbar) and in Ar/H_2 atmosphere [33].

SE in the 1.5–6.5 eV and the appropriate modeling, which was described previously, was used to estimate the a-C:H thin films thickness [8]. It was found that the thickness of the a-C:H thin films grown with: (1) $V_b > 0$ is $d = 150$ nm and (2) $V_b < 0$ is $d = 200$ nm.

In Fig. 6.10 a and b representative DNI P-h curves from the a-C:H thin films are presented. A pop-in event (step in the loading curve) is observed in the measurements of the thin films grown with $V_b < 0$ and $V_b = 20$ V. The pop-in events indicate sudden plastic deformation (possibly fracture) for normal applied load 2.1 mN and 1.7 mN, respectively [34]. It should be noted that in the case of the a-C:H thin film grown with $V_b = -11$V, the pop-in event occurred in the thin film, while that of a-C:H thin film grown with $V_b = 20$ V took place at the thin film/substrate interface. For $V_b > 20$ the P-h curves are continuous and the deformation mode is elastic/plastic (Fig. 6.10 a and b).

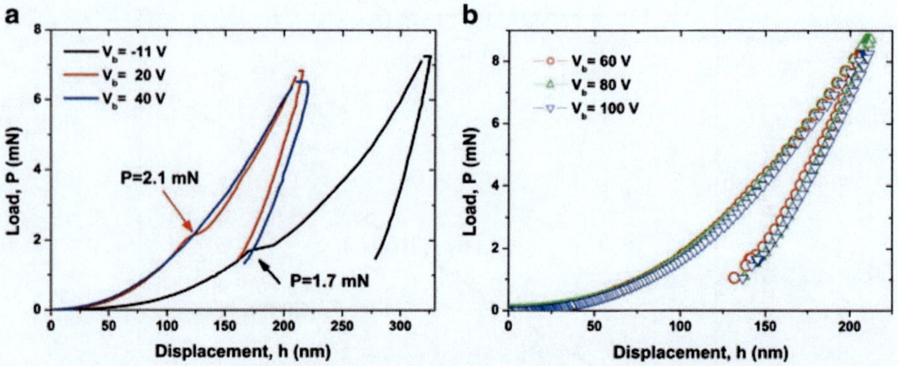

Fig. 6.10 a and b The nanoindentation load-displacement curves from the a-C:H/Si (001) developed by varying the negative applied voltage to the substrate. In Fig. 6.10 the arrows indicate the pop-in events

The appropriate modeling was used to calculate the surface E_f and H_f values of the a-C:H thin films and to separate the mechanical properties of the thin film from that of the c-Si substrate. The experimental values of E were fitted using (6.23), and those of H were fitted using (6.24) [14, 15]:

$$\left(\frac{1-\nu^2}{E}\right)_{\text{eff}} = \frac{1-\nu_f^2}{E_f}\left(1-e^{-\frac{bd}{\sqrt{\pi}h_c}}\right) + \frac{1-\nu_s^2}{E_s}e^{-\frac{bd}{\sqrt{\pi}h_c}} \tag{6.23}$$

and

$$H_{\text{eff}} = H_s + \left(H_f - H_s\right)\exp\left[-c\left(\frac{h_c}{d}\right)^2\right], \tag{6.24}$$

where ν is the Poisson ratio, h_c the contact depth, and b and c are free parameters for the fitting procedure [31, 32, 34]. E_s and H_s values of the Si(001) substrate are 160–180 GPa and 12 GPa, respectively. The subscripts s and f denote the substrate and the film, respectively, and *eff* the measured quantity. In Fig. 6.11, the calculated values of E_f and H_f are presented versus the V_b.

There is an increase of the E_f values increase from 20 GPa ($V_b < 0$) to 46 GPa ($V_b = 40$ V). Further increase of the V_b does not affect the E_f and a plateau appears in Fig. 6.11. The same behavior is also observed for H_f, which increases from 1.8 to 5.6 GPa, when $V_b \geq 40$ V. Again a value plateau appears for further increase of the V_b. The same effect of the V_b to the E_f and H_f has been also reported for several carbon based (non-hydrogenated, hydrogenated, nitrogenated) thin films [36, 37].

The elastic modulus of the a-C:H thin films was also measured using AFAM [33, 38, 39]. The AFAM is based on the principles of the contact mode atomic force microscopy with the difference that the AFAM configuration includes a piezoelectric transducer, with 2.5 MHz central frequency, which is placed below and in contact with the sample. The transducer emits longitudinal acoustic waves

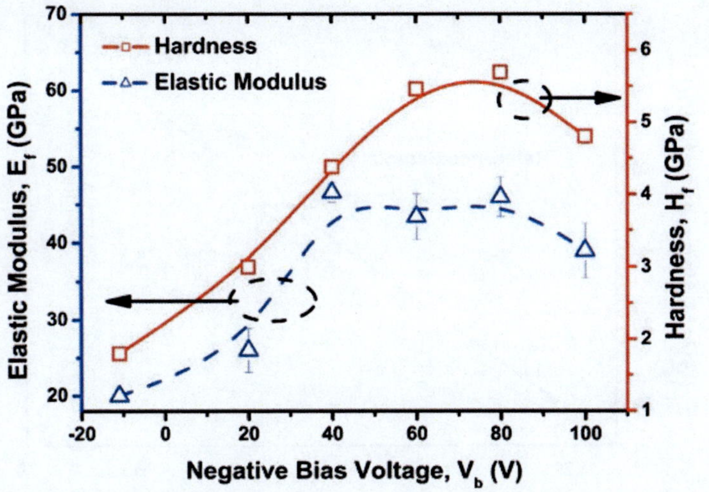

Fig. 6.11 The elastic modulus and hardness of the a-C:H thin films variation versus the applied negative bias voltage to the substrate

which cause out-of-plane vibrations of the sample surface. By setting the SPM probe (cantilever) in contact with the sample surface, we can acquire the so-called acoustic images, which represent the vibration amplitude of the cantilever. The latter vibrates in contact with the sample surface at a fixed frequency, close to the resonance one. The acoustic images of every sample is acquired simultaneously with the topographic ones, in order to study in terms of surface morphology and nanomechanical properties of the exact same surface areas of the samples. We also acquire the cantilever vibration spectra to estimate the thin films E_f [38]. The measurements were performed with a SOLVER P47H Scanning Probe Microscope (NT-MDT, NTI Instruments) in ambient environment, at room temperature. Standard silicon cantilevers with nominal spring constant $k_c = 1\,N/m$, 10 nm nominal tip radius, and 20 kHz typical resonance frequency was used for the AFAM measurements.

Rectangular cantilever, with nominal spring constant $k_c = 1\,N/m$, was selected as the scanning probe. Sufficient low contact force F_c was applied to the soft a-C:H thin films, in order to prevent damage of the thin film surface. The contact force F_c, which is equal to $F_c = k_c \Delta z$, where Δz is the static deflection of the cantilever, was set 140 nN for every scan. First, the resonant contact frequency f_c of the cantilever and the vibrated surface were measured. The contact stiffness k^* is estimated via f_c [38, 39] but for the calculation of the reduced modulus E^* a reference sample with known elastic modulus is needed. E_f is given by (25) and (26) [8, 19]:

$$E^* = E^*_{\mathrm{ref}} \frac{k^*}{k^*_{\mathrm{ref}}}, \tag{6.25}$$

$$\frac{1}{E^*} = \frac{1 - v_f}{E_f} + \frac{1 - v_t}{E_t}, \tag{6.26}$$

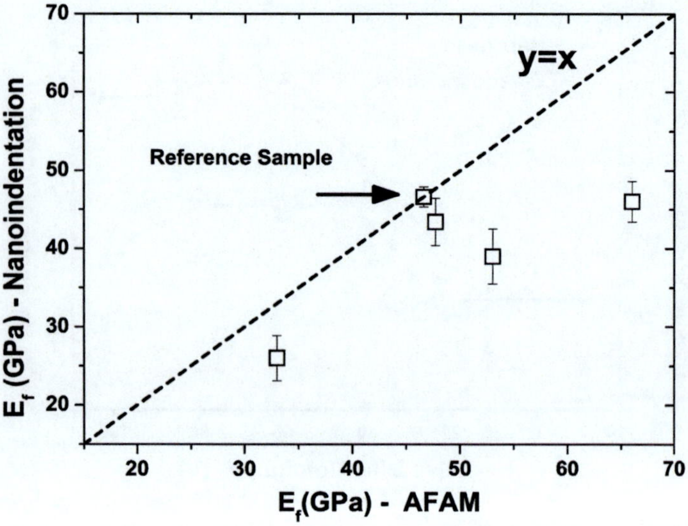

Fig. 6.12 The elastic modulus values of the a-C:H thin films measured using AFAM and Nanoindentation

where the subscripts t and ref denotes the tip and the reference sample, respectively. In Fig. 6.12, the E_f estimated via AFAM and NI is presented. As a reference for the E_f calculation via AFAM technique, the a-C:H thin film grown with $V_b = -40\,\mathrm{V}$ was used.

The same trend, an increase of E_f with decreasing V_b, is observed also for the AFAM measured E_f. In addition, the E_f values, for all the a-C:H thin films measured by the AFAM technique, appear to be higher, compared to those obtained by NI. The discrepancy probably is due to the significant difference between the forces that the probe of every technique applies to the sample, micronewton (μN) in the case of NI and nanonewton (nN) in/the case of AFAM. A similar behavior has been also reported for a-C single layer and multilayer thin films [20], in which the NI values of the E_f were found to be significantly lower with respect to the theoretically predicted ones.

Inorganic Thin Films on Flexible Polymer Substrate

Flexible polymer substrates are used in a wide range of products from food packaging products to high-tech organic optoelectronic devices (OEDs – flexible light emitting diodes, photovoltaics, displays, batteries, and fuel cells) [40]. The poor gas barrier properties of the flexible polymer substrates to the atmospheric gases affect the service life and quality of the device. For example the OEDs materials are extremely sensitive and can be easily harmed by the oxygen and

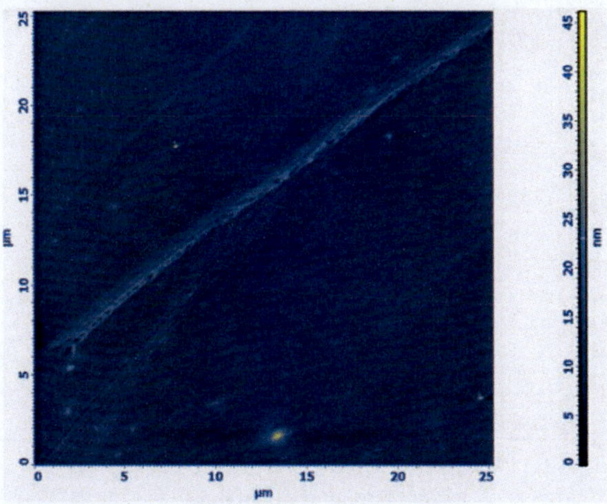

Fig. 6.13 AFM image of irregularities and defects on the flexible PET substrate

water vapors that transmit into the device and cause malfunction of the device materials/parts.

The permeation of gases is prevented by coating the polymer substrates with single layer or multilayer (for improved barrier performance) inorganic (SiOx, AlOx, SiN [41–43]), hybrid polymers (ORMOCER®, ORMOCIL [44–46]) and/or nanocomposite thin films [8]. This is an encapsulation procedure and nowadays materials and materials systems with ultrahigh barrier performance have been reported (oxygen and water vapor permeation well below $10^{-3}\,cm^3m^{-2}d^{-1}bar^{-1}$ and $10^{-3}\,gm^{-2}d^{-1}$, respectively) [47,48].

The surface morphology and roughness of the flexible polymer substrate plays a significant role in the adhesion of the barrier thin film and in the mechanical properties of the barrier thin film/polymeric substrate system [49]. Usually plasma, corona, or UV treatments are used prior to the deposition of the thin films for the minimization of the surface roughness, as well as for the activation and functionalization of the polymer surface [50–52]. In Fig. 6.13 an AFM image from the surface of the most usual flexible polymer substrate for OEDs, the PET (a 50 µm thick), is presented. The image comes from $5\,\mu m \times 5\,\mu m$ surface area, which was scanned using tapping mode. The surface root mean square (RMS) roughness is 1.7 nm. As it can be seen, the PET surface is complex and rough with many defects, (e.g., narrow pits and steep trenches). As a consequence the PET surface cannot be homogenously covered by a barrier thin film and adhesion failure can be caused [41,53].

In order to test the adhesion of barrier thin film to the flexible polymer substrate, DNI and atomic force microscopy are used. The barrier thin film was AlO_x and it was grown on 50 µm thick flexible PET substrate using *rf* MS.

In Fig. 6.14, the DNI P-h curves from AlOx/PET using $P_{max} = 532\,\mu N$ and $P_{max} = 1,239\,\mu N$ are presented. In both cases the indenter penetrated through the AlO_x barrier thin film to the PET substrate. The contact is elastic/plastic and the

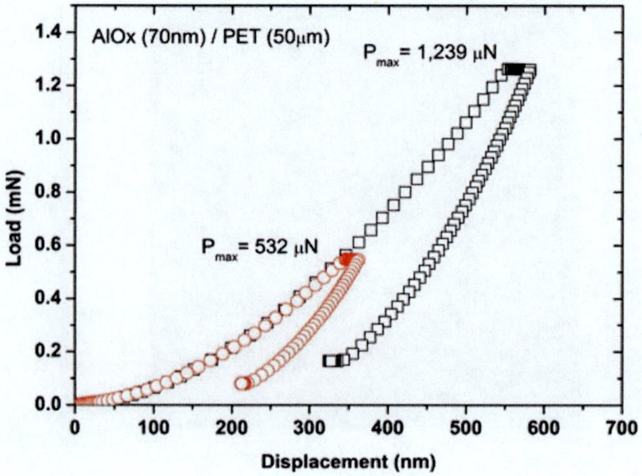

Fig. 6.14 Nanoindentation load-displacement curves for high normal applied loads

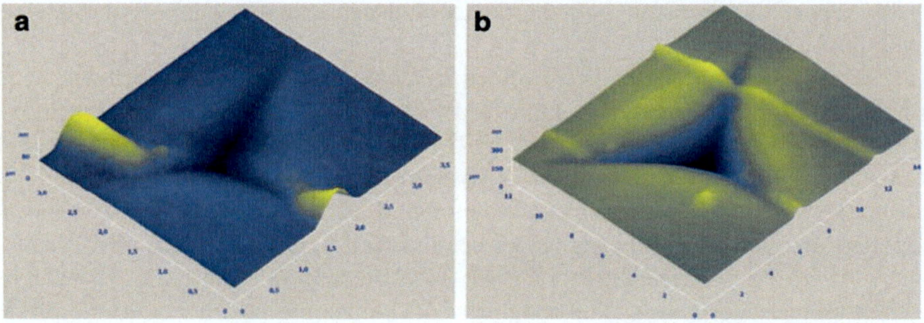

Fig. 6.15 The nanoindentation imprints on the AlOx/PET surface for maximum applied load (**a**) $P_{\max} = 532\,\mu\text{N}$ and (**b**)$P_{\max} = 1,239\,\mu\text{N}$

maximum displacement of the indenter is $h_{\max} = 339\,\text{nm}$ and $h_{\max} = 545\,\text{nm}$, respectively. Also, for 15s the $P_{\max} =$ const. and the indenter continues to penetrate into the sample, for $h = 22\,\text{nm}$ ($P_{\max} = 532\,\mu\text{N}$) and $h = 27\,\text{nm}$($P_{\max} =$ 1,239 μN) due to the NI creep deformation, which is a characteristic deformation type of the polymers [31]. In addition, the absence of any discontinuities in both the loading and the unloading portions of the P-h curves denotes that the inorganic barrier thin film remains adhere to the flexible PET substrate, although the indenter has penetrated more than 4 and 7 times the thickness of the thin film. This suggestion is also supported by imaging of the NI imprints using AFM (Fig. 6.15).

In Fig. 6.15 a and b the remained NI imprints for $P_{\max} = 532\,\mu\text{N}$ and $1,239\,\mu\text{N}$, respectively, are presented. Both imprints present the shape of the Berkovich indenter (triangular pyramid), while the deeper one appears to be four times larger in diameter from the shallower. The abrupt figures that cross the NI imprints are

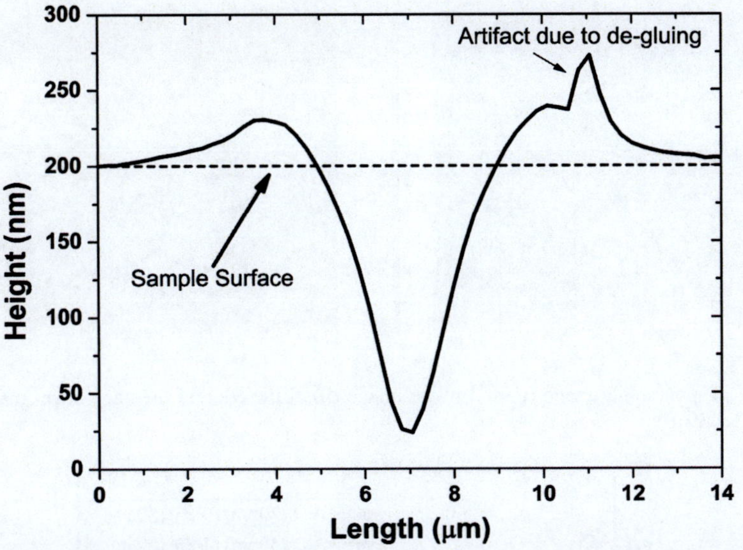

Fig. 6.16 Cross section from the deepest point of the nanoindent presented in Fig. 6.4b). The pile-up deformation at the edges of the nanoindent is presented

artifacts made during the de-gluing of the samples from the NI sample holder, in order to be measured in the AFM.

A cross section passing from the deepest point of the NI imprint (Fig. 6.16) showed the presence of a non-symmetrical pile-up of the AlOx thin film. The maximum depth of this indentation imprint is $\approx$ 200 nm, almost equal to the remaining (plastic) deformation after the unloading (Fig. 6.14 indentation for $P_{\max} \approx 532\,\mu\text{N}$). The height of the pile-up area was found to depend on the normal applied load and increased from 15–20 nm for $P_{\max} = 532\,\mu\text{N}$ to 30–40 nm for $P_{\max} = 1,239\,\mu\text{N}$. This type of mechanical deformation usually accompanies the nanoindenter penetration into inorganic bulk materials, thin films grown on inorganic substrates, and polymer substrates [31, 54]. But as it is shown in Fig. 6.15a and b the pile-up deformation also occurs in inorganic thin films, such as AlOx, grown on flexible polymeric substrate. Further studies are needed to illustrate the factors that affect the size of the pile-up area (e.g., the normal applied load, the thin film thickness) and how this deformation mode affects the cracking and the adhesion of the thin film.

Also a closer look (Fig. 6.17) at the NI imprints reveals that in both cases there is no sign of AlO_x barrier thin film de-adhesion. In Fig. 6.6 the AFM topographic and the phase images of the NI imprint made with $P_{\max} = 1,239\,\mu\text{N}$ normal applied load are presented. Especially, in the phase image the interface of the AlOx thin film and the flexible PET substrate interface is clearly seen.

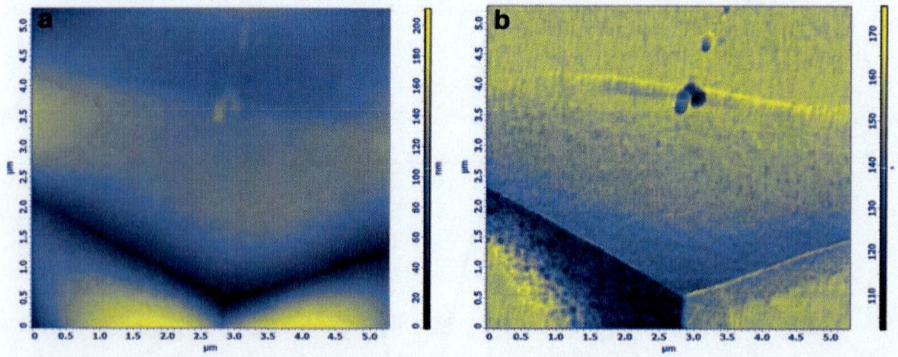

Fig. 6.17 Topographic image (**a**) and phase image (**b**) at the edge of the nanoindent made with $P_{max} = 1,239\,\mu N$

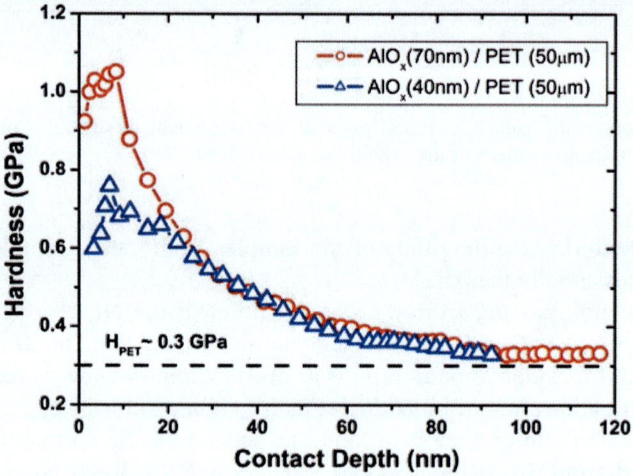

Fig. 6.18 The hardness of the AlOx/PET samples versus the contact depth

In Fig. 6.18, the hardness (H) values of two AlOx thin films, 40 nm and 70 nm thick, versus the contact depth are shown. The presented hardness values are the mean ones coming from ten indents made to every sample, with 30 μm spacing.

The thickness of the AlOx barrier thin film was found to affect the H values and the thicker film appears to be harder (for the 70 nm thick sample $H = 1.05$ GPa and for the 40 nm thick one $H = 0.76$ GPa). Both values are far from the real hardness value of the AlO_x thin film ($H_{AlOx} \approx 12$ GPa, [24]), due to the considerably lower hardness of the flexible PET substrate, which affects the measurement. ($H_{PET} =$ 0.3–0.4 GPa) [25]. This is the so-called substrate effect, which depends on the elastic and plastic behavior of both the thin film and the substrate during the NI. For the AlO_x/PET barrier thin films presented here, the substrate effect is more pronounced due to the reduced thickness of the AlOx thin films (40 and 70 nm) and to the large difference between the mechanical properties of the hard and brittle AlOx thin film and the soft, easily plastically deformed PET substrate [26, 27].

6.2.4 Wet Deposition Techniques of Thin Films

The Spin Coating method belongs to a group of methods called sol-gel and is used for thin films growth, from solution phase, on flat substrates. The principle of the method is based on an excess amount of a solution placed on the substrate, which after rotation at high speed is spread over the surface due to the centrifugal force.

The spin coating process is divided in four stages:

1. **Stage 1:** Deposition of the coating fluid onto the wafer or substrate. This can be done by using a nozzle and pouring the coating solution either static (first form a large drop on the surface and then rotate) or dynamically (drop forming during the first rotation stage). A substantial excess of coating solution is usually applied compared to the amount that is required.
2. **Stage 2:** Acceleration of the substrate up to its final, desired, rotation speed. This stage is usually characterized by aggressive fluid expulsion from the wafer surface by the rotational motion. At the end, the fluid has the adequate thickness in order to co-rotate with the surface.
3. **Stage 3:** The substrate is spinning at a constant rate and fluid viscous forces dominate fluid thinning behavior. If the liquid exhibits Newtonian viscosity and the fluid thickness is initially uniform across the wafer, then the fluid thickness profile at any following time will also be uniform.
4. **Stage 4:** The substrate is spinning at a constant rate and solvent evaporation dominates the coating thinning behavior. As the prior stage advances, the fluid thickness reaches a point where the viscosity effects yield only rather minor net fluid flow. At this point, the evaporation of any volatile solvent species will become the dominant process occurring in the coating. In fact, at this point the coating effectively "gels" because as these solvents are removed the viscosity of the remaining solution will likely rise – effectively freezing the coating in place.

Once spinning is complete, the plate is typically placed onto a hot plate (heated to temperature above the boiling of the solvent) for several seconds or minutes to initially evaporate solvent and solidify the coating.

The dry film thickness d generally is given by the empirical expression:

$$h \sim \omega^{n}, \tag{6.27}$$

where ω is the rotation speed and the n value depends on the solvent evaporation mechanism.

In case of that no evaporation occurs the dry film thickness varies with spin speed and time:

$$h \sim \omega^{-1} t^{-1} \tag{6.28}$$

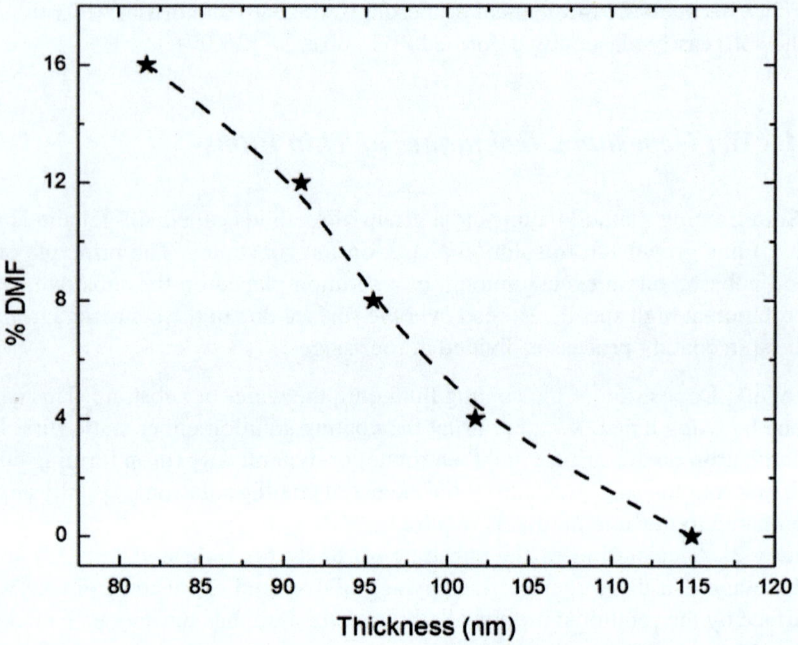

Fig. 6.19 Thin film thickness dependence from the DMF content [56]

For constant evaporation rate:

$$h \sim \omega^{-2/3} \tag{6.29}$$

In case that the evaporation rate varies (in most cases) with the square root of the spin speed:

$$h \sim \omega^{-1/2} \tag{6.30}$$

Another important parameter that affects the dry film thickness is the density of the solution before spinning that are expressed through the viscosity [55]. Thus, taking into account the kinematic viscosity that is given by the ratio of the *viscosity* (η) over the *density* (ρ) ($\nu = \eta/\rho$) the dry film thickness at constant rotation speed and spinning duration is given by:

$$h \sim p^{-1/2}. \tag{6.31}$$

This expression was verified by adding N, N-dimethylformamide (DMF) at different concentration in aqueous solutions of PEDOT:PSS and forming thin films [56]. The DMF addition resulted in a less dense solution and in the decrease of the dry film thickness as the DMF concentration is increasing (Fig. 6.19). The PEDOT:PSS film thickness was measured using SE.

6.3 Summary: Conclusion

The basic thin film deposition techniques have been briefly presented and discussed, while we focused on the MS growth of thin films and their optical and nanomechanical characterization.

The effect of the deposition conditions to the optical properties of the a-C:H thin films was studied using SE in the NIR-Vis-UV spectra range. The measured spectra was analyzed using a dispersion model based on two TL oscillators, in order to discriminate the contribution of the π–π^* and σ–σ^* interband electronic transitions that correspond to the sp^2 and sp^2/sp^3 hybridized carbon bonds, respectively. The increase of the applied negative bias voltage during the a-C:H thin film growth was found to increase the sp^3 and sp^2 hybridized carbon bonds fractions in the film as well as to lead to the formation of more transparent a-C:H thin films (lower band gap).

The nanomechanical properties of hard and soft carbon based thin films developed on Si (001) substrate in single layer and multilayer form was discussed. It was proved that the multilayer structure affects the nanomechanical properties and leads to the development of harder thin films compared to the single layer ones.

In the case of the soft a-C:H thin films the applied negative bias voltage affects their nanomechanical behavior. The harder and more coherent a-C:H thin films were grown by increasing the V_b. This effect of the V_b to the elastic modulus of the a-C:H thin films was also confirmed by AFAM, while there was a difference between the E_f values measured by NI and AFAM.

Also the case of an inorganic thin film (AlO_x) grown on a flexible polymer (PET) thin film was presented. The effect of the NI deformation to the AlO_x/PET samples was studied using AFM. It was found that the AlOx thin film remained adhere to the PET substrate although the nanoindenter maximum penetration depth exceeded seven times the thickness of the AlO_x thin film. Also the AFM images showed the presence of pile-up regions that surrounds the NI imprint. The height of the pile-up was found to depend on the maximum normal applied load.

Finally, the Spin Coating thin film deposition technique from the liquid phase is presented and the effect of the polar solvent addition to the solution to the PEDOT:PSS thickness is discussed.

References

1. D.M. Mattox, *Handbook of Physical Vapor Deposition (PVD) Processing* (Noyes Publications, Westwood, 1998)
2. M. Ohring, *The Material Science of Thin Films* (Academic, San Diego, 1992)
3. D.M. Dobkin, M.K. Zuraw, *Principles of Chemical Vapor Deposition* (Kluwer, New York, 2003)
4. P. Sigmund, *"Sputtering by Ion Bombardment: Theoretical Concepts" in "Topics in Applied Physics* (Springer, New York, 1981.

5. A. Belkind, A. Freilich, J. Lopez, Z. Zhao, W. Zhu, K. Becker, New J. Phys. **7** 90 (2005)
6. K. Sarakinos, J. Alami, S. Konstantinidis, Surf. Coat. Technol. **204** 1661–1684 (2010)
7. V. Kouznetsov, K. Macak, J.M. Schneider, U. Helmersson, I. Petrov, Surf. Coat. Technol. **122** 290 (1999)
8. S. Logothetidis, *Handbook of Thin Film Technology*, Vol. 2, ed. by H.S. Nalwa (Academic, San Diego, (2002), p. 277
9. R. Azzam, N. Bashara, *Ellipsometry and Polarized Light* (North-Holland, Amsterdam, 1977)
10. S. Logothetidis, (2002) *Handbook of Thin Film Technology*, Vol. 2, ed. by H.S. Nalwa, R. Azzam, N. Bashara. Ellipsometry and Polarized Light, (Academic Press, Amsterdam, 1977), p. 277
11. G. Irene, H. Tompkins, *Handbook of Ellipsometry* (William Andrew Publishing, Norwich, 2005)
12. A. Laskarakis, S. Logothetidis, M. Gioti, Phys. Rev. B **64** 125419 (2001)
13. G. Jellison, F. Modine, Appl. Phys. Lett. **69** 371 (1996)
14. G. Jellison, Thin Solid Films **313–314** 33 (1998)
15. J. Hong, S. Lee, C. Cardinaud, G. Turban, J. Non Cryst. Solids **265** 125 (2000)
16. Z. Chen, Y. Yu, J. Zhao, C. Ren, X. Ding, T. Shi, X. Liu, Surf. Coat. Technol. **104** 380 (1998)
17. A. Tibrewala, E. Peiner, R. Bandorf, S. Biehl, H. Lu, Appl. Surf. Sci. **252** 5387 (2006)
18. J. Hong, A. Goullet, G. Turban, Thin Solid Films **364** 144 (2000)
19. M. Shinohara, K. Cho, H. Shibata, K. Okamoto, T. Nakatani, Y. Matsuda, H. Fujiyama, Thin Solid Films **516** 4379 (2008)
20. S. Kassavetis, P. Patsalas, S. Logothetidis, J. Robertson, S. Kennou, Diam. Relat. Mater. **16** 1813 (2007)
21. J. Zhu, J. Han, X. Han, S. Meng, A. Liu, X. He, Opt. Mater. **28** 473 (2006)
22. S. Logothetidis, Diam. Relat. Mater. **12** 141 (2003)
23. M. Gioti, S. Logothetidis, Diam. Relat. Mater. **12** 957 (2003)
24. D. Tabor, *The Hardness of Metals* (Clarendon Press, Oxford, 1951)
25. W.C. Oliver, G.M. Pharr, J. Mater. Res. **7** 1564 (1992)
26. A.C. Fischer-Cripps, *Nanoindentation* (Springer, New York, 2004)
27. I.N. Sneddon, Int. J. Eng. Sci. **3** 47 (1965)
28. J.B. Pethica, W.C. Oliver, Phys. Scr. T **T19** 61 (1987)
29. W.C. Oliver, G.M. Pharr, MRS Bull. **35** 897 (2010)
30. W.C. Oliver, G.M. Pharr, J. Mater. Res.**19** 3 (2004)
31. S. Logothetidis, S. Kassavetis, C. Charitidis, Y. Panayiotatos, A. Laskarakis, Carbon **42** 1133 (2004)
32. A.K. Bhattacharya, W.D. Nix, Int. J. Solids Struct. **24** 1287 (1988)
33. S. Kassavetis, S. Logothetidis, G.M., Matenoglou, Surf. Coat. Technol. **200** 6400 (2006)
34. J. Qi, K.H. Lai, C.S. Lee, I. Bello, S.T. Lee, J.B. Luo, S.Z. Wen, Diam. Relat. Mater. **10** 1833 (2001)
35. M.F. Doerner, W.D. Nix, J. Mater. Res. **1** 601 (1986)
36. J. Robertson, Mater. Sci. Eng., R Rep. **37** 129 (2002).
37. A. Laskarakis, S. Logothetidis, C. Charitidis, M. Gioti, Y. Panayiotatos, M. Handrea, W. Kautek, Diam. Relat. Mater. **10** 1179 (2001)
38. U. Rabe, S. Amelio, E. Kester, V. Scherer, S. Hirsekorn, W. Arnold, Ultrasonics **38** 430 (2000)
39. U. Rabe, M. Kopycinska, S. Hirsekorn, J. Muñoz Saldaña, G.A. Schneider, W. Arnold, J. Phys. D: Appl. Phys. **35** 2621 (2002)
40. S. Logothetidis, A. Laskarakis, Eur. Phys. J. Appl. Phys. **46** 12502 (2009)
41. Y. Leterrier, Prog. Mater. Sci. **48** 1 (2003)
42. P.F. Carcia, R.S. McLean, M.H. Reilly, M.D. Groner, S.M. George, Appl. Phys. Lett. **89** 031915 (2006)
43. S. da Silva Sobrinho, M. Latreche, G. Czeremuszkin, J.E. Klemberg-Sapieha, M.R. Wertheimer, J. Vac. Sci. Technol. A **16** 3190 (1998)
44. K.H. Haas, S. Ambergschwab, K. Rose, G. Schottner, Surf. Coat. Technol. **111**, 72 (1999)
45. S.Y. Lee, J.D. Lee, S.M. Yang, J. Mater. Sci. **34** 1233 (1999)

46. D. Georgiou, A. Laskarakis, S. Logothetidis, S. Amberg-Scwhab, U. Weber, M. Schmidt, K. Noller, Appl. Surf. Sci. **255** 8023 (2009)
47. A. Laskarakis, S. Logothetidis, D. Georgiou, S. Amberg-Schwab, U. Weber, Thin Solid Films **517** 6275 (2009)
48. C. Charton, N. Schiller, M. Fahland, A. Hollander, A. Wedel, K. Noller, Thin Solid Films **502** 99 (2006)
49. S. Yang, Y.W. Zhang, K. Zeng, J. Appl. Phys. **95** 3655 (2004)
50. K.W. McElhaney, J.J. Vlassak, W.D. Nix, J. Mater. Res, **13** 1300 (1998)
51. S. Kassavetis, K. Mitsakakis, S. Logothetidis, Mater. Sci. Eng. C **27** 1456 (2007)
52. R. Saha, W.D. Nix, Acta Mater. **50** 23 (2002)
53. S. Ben Amor, G. Baud, J.P. Besse, M. Jacquet, Thin Solid Films **293** 163 (1997)
54. U.A. Handge, Y. Leterrier, G. Rochat, I.M. Sokolov, A. Blumen, Phys. Rev. E **62** 7807 (2000)
55. D. Meyerhofer, J. Appl. Phys. **49** 3993 (1978)
56. C. Gravalidis, A. Laskarakis, S. Logothetidis, EPJ Appl. Phys. **46** 12505 (2009)

Chapter 7
Implementation of Optical Characterization for Flexible Organic Electronics Applications

A. Laskarakis and S. Logothetidis

Abstract One of the most rapidly evolving sectors of the modern science and technology is the flexible organic electronic devices (FEDs) that are expected to significantly improve and revolutionize our everyday life. The FED application includes the generation of electricity by renewable sources (by organic photovoltaic cells – OPVs), power storage (thin film batteries), the visualization of information (by organic displays), the working and living environment (ambient lighting, sensors), safety, market (smart labels, radio frequency identification tags – RFID), textiles (smart fabrics with embedded display and sensor capabilities), as well as healthcare (smart sensors for vital sign monitoring), etc. Although there has been important progresses in inorganic-based Si devices, there are numerous advances in the organic (semiconducting, conducting), inorganic, and hybrid (organic–inorganic) materials that exhibit desirable properties and stability, and in the synthesis and preparation methods. The understanding of the organic material properties can lead to the fast progress of the functionality and performance of FEDs. The investigation of the optical properties of these materials can promote the understanding of the optical, electrical, structural properties of organic semiconductors and electrodes and can contribute to the optimization of the synthesis process and the tuning of their structure and morphology. In this chapter, we will describe briefly some of the advances toward the implementation of optical characterization methods, such as Spectroscopic Ellipsometry (SE) from the infrared to the visible and ultraviolet spectral region for the study of materials (flexible polymer substrates, barrier layers, transparent electrodes) to be used for application in the fabrication of FEDs.

A. Laskarakis (✉) · S. Logothetidis
Department of Physics, Laboratory for Thin Films-Nanosystems and Nanometrology (LTFN), Aristotle University of Thessaloniki, GR-54124 Thessaloniki, Greece
e-mail: alask@physics.auth.gr

S. Logothetidis (ed.), *Nanostructured Materials and Their Applications*, NanoScience and Technology, DOI 10.1007/978-3-642-22227-6_7,

7.1 Introduction

Organic electronics is one of the most rapidly emerging sectors of the modern science and technology and it is expected to improve our everyday life in the near future [1–4]. The advances in materials and processes for organic electronics are expected to revolutionize existing applications and will create novel applications, such as flexible displays (organic light-emitting diodes – OLEDs) for visualization of information and lighting, organic photovoltaic cells (OPVs), organic thin film transistors (OTFTs) embedded onto flexible organic circuits, flexible thin film batteries, flexible organic sensors and biosensors, as well as radio frequency identification tags (RFID) [1–8]. Also, new applications are expected in several sectors, such as in automotive industry, architecture, building construction, healthcare, etc. [3].

The most intriguing benefits from the use of organic semiconducting materials for deposition onto flexible polymeric substrates include the low cost of the production processes, the mechanical flexibility of the device, the ability to be rolled when the device is not used, the light weight and conformable design [1–3]. These make this type of devices most attractive for large-scale and mobile applications. Moreover, entirely new design concepts for consumer electronics have emerged, since flexible organic electronic devices (FEDs) can be adopted to follow complex surface shapes [1–3].

The field of flexible organic electronics includes several innovations in materials, such as organic semiconductors (small molecule and polymers) with sufficient electrical conductivity, transparent organic electrodes that can be deposited by printing processes, hybrid barrier materials for the protection of the sensitive device materials from atmospheric gas permeation, and finally, flexible polymeric substrates, such as PolyEthylene Terephthalate (PET), and Polyethylene naphthalate (PEN) that will replace the Si and glass rigid substrates [9–11].

The performance, efficiency, and lifetime of FEDs are defined by the physical (optical, electrical, structural, mechanical) properties and the nanoscale morphology of the organic semiconductor, electrode, and barrier materials [2]. These should meet specific and advanced requirements, such as high optical transparency, high electrical conductivity, structural stability, ultra low atmospheric gas permeability, film–substrate adhesion, etc. The knowledge of the optical properties of these materials in the infrared-ultraviolet spectral region is of considerable importance and it can contribute to the understanding of the bonding and electronic structure and microstructure, as well as on the optical transparency and structure-property relationships of these materials [1].

In situ and real-time Spectroscopic Ellipsometry (SE) is a powerful, non-destructive, and surface sensitive optical technique that has been used for the investigation of optical vibrational, structural, and morphological properties, the composition and the growth mechanisms of inorganic and organic bulk materials and thin films [11, 12]. The implementation of real-time SE monitoring and control to large-scale production of functional thin films for numerous applications, will lead to the optimization of the materials quality and increase in production yield [13, 14].

In this chapter, we will provide an overview of the implementation of SE on the optical characterization of state-of-the-art materials (flexible polymer substrates, barrier layers, inorganic and organic transparent electrodes) for the fabrication of FEDs. These will prove the potentiality of SE for implementation to: (a) lab scale processes as an important research tool that will contribute to the understanding and the optimization of the materials properties, and (b) industrial scale processes as a powerful quality control tool for the roll-to-roll (r2r) fabrication of FEDs.

7.2 Optical Characterization of Materials

SE is a very powerful technique for the characterization of bulk materials, surfaces, nanomaterials, and thin film systems [11, 12, 15–17]. Under the appropriate circumstances, SE can determine the thickness and the optical properties of materials, more accurately than any other technique. In addition, SE can provide information on the optical functions, surface roughness, interface layers as well as the volume fraction of multiple phases in composite bulk materials and thin films. SE can be implemented for the detailed characterization of all types of materials, such as dielectrics, semiconductors, metals, organic materials, opaque, semitransparent, or even transparent materials [12, 17]. This technique and its instrumentation relies on the fact that the reflection of a dielectric interface depends on the polarization of the light while the transmission of light through a transparent layer changes the phase of the incoming wave depending on the refractive index of the material. Applications include the accurate thickness measurement of thin films, the identification of the optical properties of materials and thin films, the evaluation of vibrational, electronic, compositional and nanostructural properties, and the optical characterization of surfaces and interfaces [11, 12, 15].

SE measures the optical response of bulk materials and thin films as a function of the photon energy ω [11, 12, 15]. The determined properties are the complex refractive index $\tilde{n}(\omega)$ and the complex dielectric function $\tilde{\varepsilon}(\omega)$. The complex refractive index is related to dispersion and absorption of light by the following relation [11, 12]:

$$\tilde{n}(\omega) = n(\omega) + ik(\omega) \tag{7.1}$$

The complex dielectric function $\tilde{\varepsilon}(\omega)$ is directly related to the material properties, and is connected to the refractive index through the equation:

$$\tilde{\varepsilon}(\omega) = \varepsilon_1 + i\varepsilon_2 \equiv \tilde{n}^2(\omega) = (n + i\kappa)^2 \Rightarrow \begin{cases} \varepsilon_1 = n^2 - k^2 \\ \varepsilon_2 = 2nk \end{cases} \tag{7.2}$$

In the case of the reflection of a light beam at the planar interface between two semi-infinite optically isotropic media 0 and 1, when the light beam does not penetrate the medium (1), either due to its high absorption coefficient or its infinite thickness, we are referred to a two-phase (ambient-substrate) system, or a bulk material surrounding by medium (0). In this case the ratio of the p–, s– Fresnel reflection coefficients, namely the complex reflection ratio is the quantity measured directly by SE and it is given by the expression [11, 12, 15]:

$$\tilde{\rho} = \frac{\tilde{r}_p}{\tilde{r}_s} = \left|\frac{\tilde{r}_p}{\tilde{r}_s}\right| e^{i(\delta_p - \delta_s)} = \tan \Psi e^{i\Delta}. \tag{7.3}$$

In this expression, Ψ and Δ are the ellipsometric angles, and for a bulk material take values $0° < \Psi < 45°$ and $0° < \Delta < 180°$. From an SE measurement, the complex reflection ratio $\tilde{\rho}$ is estimated, through the calculation of amplitude ratio $\tan\Psi$ and the phase difference Δ. From these two quantities one can extract all the other optical constants of the material. For example, the complex dielectric function of a bulk material with smooth surfaces is directly calculated by the following expression [11, 12, 15]:

$$\tilde{\varepsilon}(\omega) = \varepsilon_1 + i\varepsilon_2 = \tilde{n}^2(\omega) = \tilde{\varepsilon}_0 \sin^2\theta_0 \left\{ 1 + \left[\frac{1 - \tilde{\rho}(\omega)}{1 + \tilde{\rho}(\omega)}\right]^2 \tan^2\theta_0 \right\}, \tag{7.4}$$

where θ_0 is the angle of incidence of the beam, and $\tilde{\varepsilon}_0$ the dielectric constant of the ambient medium (for the case of air $\tilde{\varepsilon}_0 = 1$).

In the case of a film that is grown on a bulk substrate (see Fig. 7.1), we have the three-phase model where the film with thickness d [medium (1)] is confined between the semi-infinite ambient [medium (0)] and the substrate [medium (2)]. The complex reflectance ratio is defined as [11, 12, 15]:

$$\tilde{\rho} = \frac{\tilde{R}_p}{\tilde{R}_s} \tag{7.5}$$

$$\tilde{R}_p = \frac{\tilde{r}_{01p} + \tilde{r}_{12p} e^{i2\beta}}{1 + \tilde{r}_{01p}\tilde{r}_{12p} e^{i2\beta}}, \tag{7.6}$$

$$\tilde{R}_s = \frac{\tilde{r}_{01s} + \tilde{r}_{12s} e^{i2\beta}}{1 + \tilde{r}_{01s}\tilde{r}_{12s} e^{i2\beta}}, \tag{7.7}$$

where, $\tilde{r}_{01i}$ and $\tilde{r}_{12i}$ $(i = p, s)$ are the *Fresnel* reflection coefficients for the interfaces between medium (0) and (1), and medium (1) and (2), respectively. These depend on film thickness due to the multiple reflections of light between the media (0)–(1) and (1)–(2), through the phase angle β given by [11, 12, 15]:

$$\beta = 2\pi(d/\lambda)\left(n_1^2 - n_0^2 \sin^2\theta_0\right)^{1/2}, \tag{7.8}$$

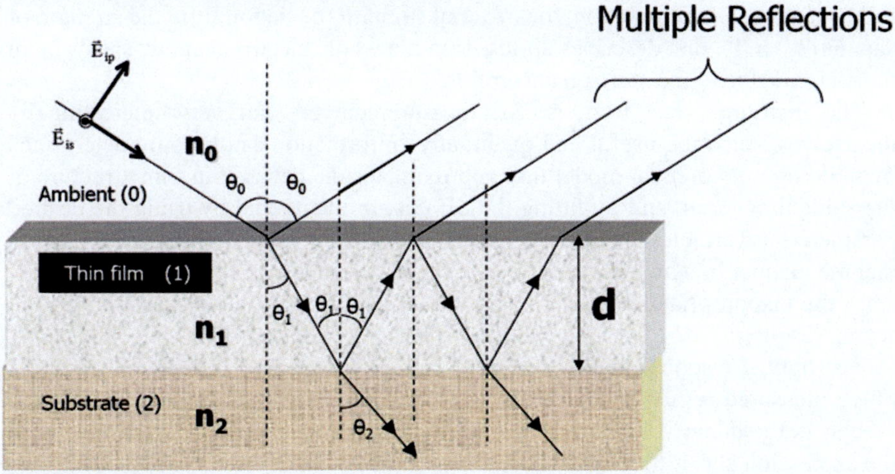

Fig. 7.1 Representation of the light reflection on the surface of a thin film deposited onto a bulk substrate

where λ is the wavelength, θ_0 is the angle of incidence and n_0, n_1 are the complex refraction indices of the ambient and the film, respectively. Thus, the measured quantity is the pseudodielectric function $\langle\varepsilon(\omega)\rangle = \langle\varepsilon_1(\omega)\rangle + i\langle\varepsilon_2(\omega)\rangle$ which contains information also about the substrate and the film thickness. As a result, the phase angle β diminishes in the energy region of high absorption, leading to $\tilde{R}_i = \tilde{r}_{01i} = \tilde{r}_i$, $(i = p, s)$ and consequently to $\langle\rho\rangle = \rho$. Thus, at the absorption bands we obtain information only of the optical properties of the bulk of the film.

The real $\varepsilon_1(\omega)$ and the imaginary $\varepsilon_2(\omega)$ parts of the dielectric function are strongly related through the well-known Kramers–Kronig relation (is based on the principle of the causality):

$$\varepsilon_1(\omega) = 1 + \frac{2}{\pi} P \int_0^{\infty} \frac{\omega' \varepsilon_2(\omega')}{\omega'^2 - \omega^2} d\omega' \tag{7.9}$$

$$\varepsilon_2(\omega) = -\frac{2\omega}{\pi} P \int_0^{\infty} \frac{\varepsilon_1(\omega') - 1}{\omega'^2 - \omega^2} d\omega', \tag{7.10}$$

where, P means the principal value of the integral around the characteristic of the material electronic resonance $(\omega' = \omega)$ and

$$\varepsilon_1(\omega = 0) = 1 + \frac{2}{\pi} P \int_0^{\infty} \frac{\varepsilon_2(\omega')}{\omega'} d\omega' \tag{7.11}$$

is the static dielectric function, the material strength (deviation from the strength of vacuum $\varepsilon_0 = 1$), that describes all losses in the whole electromagnetic spectrum in the material due to the electron absorption.

The measured data from an SE measurement are not very interesting by themselves, since the useful and quantitative information can be only determined by building a theoretical model that approximates the actual thin film structure of the studied material, and by fitting the SE data to this model by using the desired parameters as variables in the numerical analysis [12,17,18]. It has to be emphasized that the manner in which the data analysis is performed is of significant importance since the inappropriate modeling of the measured SE parameters can often lead to wrong results.

The optical response of the thin films can be deduced by the parameterization of the measured $\langle \varepsilon(\omega) \rangle$ by the use of appropriate theoretical models. One of these models is the damped harmonic oscillator (Lorenz model), which is described by the expression [11, 12, 17, 18]:

$$\tilde{\varepsilon}(\omega) = 1 + \frac{\mathrm{f}\omega_0^2}{\omega_0^2 - \omega^2 + i\Gamma\omega}, \tag{7.12}$$

where ω is the energy of light and ω_0 is the absorption energy of the electronic transition. The constants f and Γ denote the oscillator strength and the damping (broadening) of the specific transition, respectively. The quantity $\varepsilon_1(\omega = 0)$ is given by the relation $\varepsilon_1(\omega = 0) = 1 + \omega_p^2/\omega_0^2 = \mathrm{f}$, which is the static dielectric constant and represents the contribution of the electronic transition that occurs at an energy ω_0 in the NIR-Visible-UV energy region, on the dielectric function and ω_p is the plasma energy [19,20].

In the case that more than one electronic transition occurs, their contribution in $\varepsilon_1(\omega = 0)$ is accounted by the summation $\varepsilon_1(\omega = 0) = 1 + \Sigma f_i$ where Σf_i describes the losses in the material in the whole electromagnetic region due to the electronic transitions. In the case of semiconducting materials, interband electronic transitions take place due to the interaction between the electromagnetic radiation (photons) and the matter (electrons). According to classical Lorentz oscillator model, the dielectric function is given by (7.12). However, one of the models that are used for the modeling of the measured dielectric spectra of amorphous semiconductors is the Tauc–Lorentz (TL) model [14, 21, 22]. This is based on the combination of the classical Lorentz dispersion relation and the Tauc density of states in the proximity of the fundamental optical gap ω_g. This results to an asymmetrical Lorentzian lineshape for the imaginary part $\varepsilon_2(\omega)$ of the dielectric function. The TL dispersion model is described by the following relations in which the real part $\varepsilon_1(\omega)$ is determined by the imaginary part $\varepsilon_2(\omega)$ by the Kramers–Kronig integration (7.13) [14,21,22]:

$$\varepsilon_2(\omega) = \begin{cases} \dfrac{\mathrm{A}\omega_0\mathrm{C}(\omega - \omega_\mathrm{g})^2}{(\omega^2 - \omega_0^2)^2 + \mathrm{C}^2\omega^2}\dfrac{1}{\omega} & \omega > \omega_g \\ 0 & \omega \leq \omega_g \end{cases} \tag{7.13}$$

where ω_g is the fundamental band gap energy, A is related to the transition probability, ω_0 is the Lorentz resonant energy, and C the broadening term, which is a measure of the materials disorder [14, 21, 22].

In the following, we present some representative examples of the investigation of the optical properties by SE of different inorganic, organic, and polymer materials such as flexible polymeric substrates, organic semiconductors, transparent conductive oxides (inorganic and organic) that are implemented for the fabrication of FEDs.

7.3 Flexible Organic Electronic Devices

Although there are several examples of FEDs, among the most important applications are the OPVs and OLEDs. The OPVs will be used for solar energy harvesting and conversion to electric power for supplying the electric grid for consumer use. It is expected that the 1% of the global energy will be generated by OPVs until 2015 whereas this fraction will reach 10% until 2020. However, their cost should be reduced by a factor of 10 and their efficiency should reach the value of the inorganic PVs (~22.7% on 780 cm^2 area for crystalline Si) [1].

The most efficient and widely known type of OPV device is the bulk heterojunction [1–3]. This consists of two organic materials (an electron acceptor and a polymer) sandwitched between the anode and the cathode. The final OPV structure is consisted of the flexible polymeric substrate, the transparent barrier layers and the transparent electrodes (Fig. 7.2a). The state-of-the-art materials for OPVs are the blends of poly(p-phenylene vinylene) derivatives (PPVs) or (poly(3-hexylthiophene) (P3HT) (for exciton generation & hole transport) and the organo-soluble fullerene derivative (PCBM) (for electron transport) [4]. These

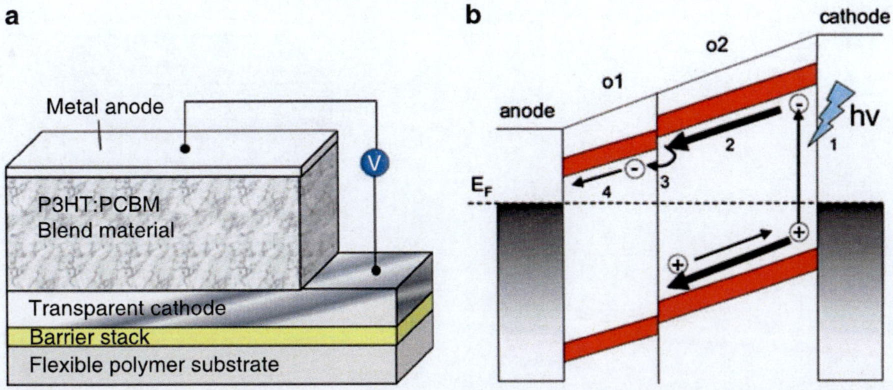

Fig. 7.2 (**a**) Schematic representation of an OPV device structure, (**b**) The four-step operation principle of a bilayer OPV device

conjugated molecules are electronically active because of their highly polarizable π-systems [4].

During the absorption of a photon, an electron is excited from the highest occupied molecular orbital (HOMO) to the lowest unoccupied molecular orbital (LUMO) (Fig. 7.2b). This electron-hole pair (exciton) relaxes with a binding energy between 0.1 and 1.4 eV (in inorganic semiconductors this is a few meV). The bound excitons must migrate to an interface where there is a sufficient chemical potential energy drop in order to drive dissociation into an electron-hole pair that spans the interface across the donor (material with low electron affinity) and acceptor (high electron affinity) [1–3, 5]. After dissociation, each charge carrier must avoid traps and recombination and reach the appropriate contact while avoiding traps and recombination. The control of the blend morphology, in combination with broad band light absorption, long live excited states and high charge carrier mobilities are essential in order to achieve high efficiency [1–3, 23].

Another major application of FEDs is the flexible OLEDs. A schematic energy-level diagram of a multilayer OLED can be seen in Fig. 7.3. In OLEDs, the active organic (electroluminescent-EL) layer is formed in the middle of a high work-function (φ1) anode and a low-work-function (φ2) cathode. The anode materials should allow easy hole injection and it is consisted by transparent conductive oxides (TCOs) such as Indium Tin Oxide (ITO) (φ = 4.7 eV) or Poly(3,4-ethylenedioxythiophene) poly(styrenesul-fonate) (PEDOT:PSS) [1,2]. The cathode should allow easy electron injection and it must have band gap values higher than the one of the TCO. It is consisted of metals such as Ca, Mg, and Cd, whereas the typical metals such as Ag and Al have work function values of 5.1 and 4.3 eV, respectively. By applying an external driving voltage, electrons are injected into the conduction band and holes into the valence band of a semiconducting polymer. Upon injection from the electrodes, electrons, and holes self-localize to form negative and positive polarons, which travel under the apparent electric field in opposite directions. When two oppositely charged polarons meet, they can form bound electron-hole pairs (excitons), which produce photons [3].

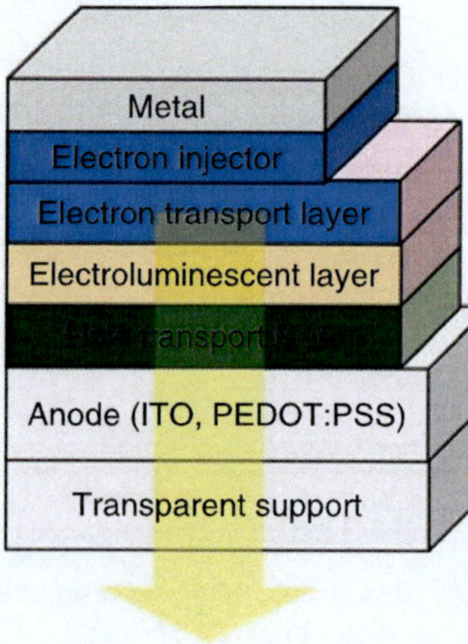

Fig. 7.3 Schematic representation of an OLED device structure

Despite the continuous advances in the materials and processes for flexible organic electronics materials, there are several open issues that have to be addressed, including the understanding of the mechanisms for the blend morphology of the organic semiconductor, the increase of the charge mobility of the transparent electrodes and the optimization of the barrier response of the barrier materials that are used for the encapsulation of the device.

7.4 Results and Discussion

7.4.1 Flexible Polymeric Substrates

The most important flexible polymers that are used as substrates for the deposition of organic semiconducting thin films are the PET and PEN [10, 24, 25]. These have attracted significant attention due to their use in several applications and industrial fields, as for the fabrication of high performance polarization optics, data storage and recording media and optoelectronic devices, in food and pharmaceutical packaging, as well as in artificial heart valves, sutures, and artificial vascular grafts [10,24]. Moreover, PET and PEN can be incorporated to large-scale manufacturing processes, since they exhibit a combination of very important properties such as easy processing, flexibility, low cost, good mechanical properties and reasonably high resistance to oxygen and water vapor penetration. In addition, these can be produced in the form of web rolls of several centimetres width and of several meters to kilometres of length, allowing the production of organic electronic devices by large-scale r2r processes [1, 10, 24, 26].

The monomer units of PET and PEN are schematically shown in Fig. 7.4. The unit cell of PET, which is triclinic with a density of $1.455\,g/cm^3$, presents a C_{2h} point symmetry, and consists of an aromatic ring and an ester function that form the terephthalate group, and by a short aliphatic chain that constitutes the ethylene segment. PEN exhibits large similarities with PET and it has a triclinic unit cell with the addition of a second phenyl ring, forming the naphthalene group [9, 24].

The large-scale production of these polymeric films includes a stretching process in order to achieve the necessary thickness and other desirable mechanical properties. Although the unit cells of PET and PEN have a triclinic crystal structure, the applied mechanical stretching process leads in the preferential orientation of the macromolecular chains toward the stretching direction (referred as Machine Direction – MD). This induces an optical anisotropy of the PET and PEN films. In this way, the polymer films consist of oriented crystalline-like regions embedded in a non-oriented non-crystalline matrix, where the macromolecular chains are randomly mixed. During the investigation of their optical properties by SE, the spectra include the averaged information from the dielectric response of both regions (oriented and non-oriented) [1,10,24,26]. Therefore the polymer films can be regarded as effective dielectric media, composed of oriented and non-oriented regions, and they can be represented by a general dielectric tensor carrying information about the optical

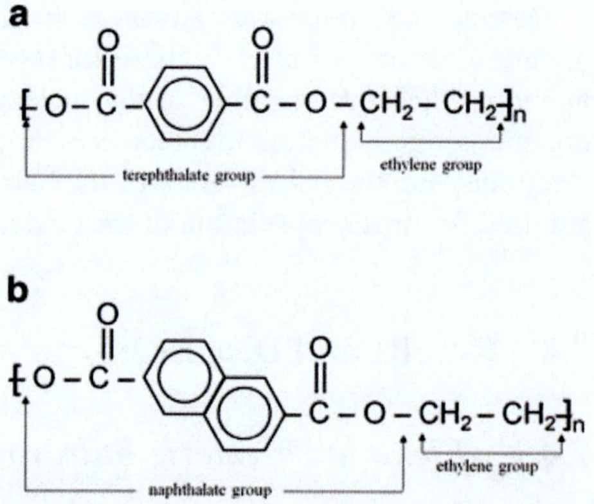

Fig. 7.4 Monomer unit of (**a**) PET and (**b**) PEN materials

response of the films and about the orientation of the macromolecular PET and PEN chains with respect to the MD [1, 10, 24, 26].

The pseudodielectric function $\langle\varepsilon(\omega)\rangle$ of the PET and PEN polymeric substrates in the Vis-fUV spectral region are shown in Fig. 7.5. Both the real and imaginary parts of the measured $\langle\varepsilon(\omega)\rangle$ at energies below 4 and 3.5 eV for PET and PEN, respectively, is consisted of interference fringes. This is attributed to the multiple reflections of light at their back interface, as a result of the optical transparency that these polymer exhibit. At higher photon energies, the optical absorption of the various electronic transitions takes place at specific characteristic energies.

More specifically, in the case of PET (Fig. 7.5a), the absorption peaks I and II can be attributed to the $n \rightarrow \pi^*$ electronic transition of the non-bonded electron of the carbonyl O atom. The peak III (PET), which is possibly attributed to the spin-allowed, orbitally-forbidden $^1A_{1g} \rightarrow ^1 B_{1u}$ transition, has been reported to be composed by two sub-peaks with parallel polarization dependence [10, 24, 25]. In the case of PEN, the existence of naphthalene group leads to the significant shift of the absorption bands of PEN to lower energies and to a characteristic split in all of them than in PET. Therefore, in PEN, the peak III can be decomposed to the sub-peaks IIIa (4.16 eV), IIIb, (4.32 eV), IIIc (4.49 eV) [10, 24, 25]. Finally, peak IV (PET) can be analyzed to two sub-peaks with different polarization (6.33 and 6.44 eV) after molecular orbital calculation based on π electron approximation and it can be attributed to the $^1A_{1g} \rightarrow ^1 B_{1u}$ electronic transition of the para-substituted benzene and naphthanene rings of the PET and PEN films with polarization rules rings plane. This peak in PEN can be analyzed to three sub-peaks at 4.97(IVa), 5.2(IVb), and 5.7(IVc) eV [10, 24, 25].

Also, the optical response of PET and PEN polymeric films in the IR spectral region are shown in Fig. 7.6. Between 900 and 1,800 cm^{-1} the strong absorption bands show the contribution of the vibrational modes corresponding to the IR-active chemical bonds of PET and PEN. Above 1,800 cm^{-1}, both films are optically transparent and their Fourier Transform IR Spectroscopic Ellipsometry (FTIRSE)

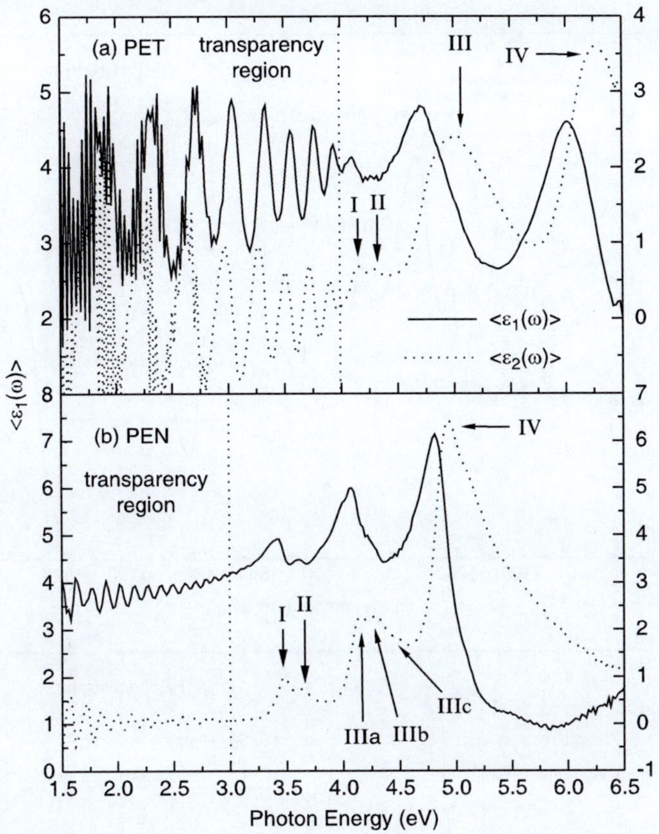

Fig. 7.5 Measured $\langle\varepsilon(\omega)\rangle$ of (**a**) PET and (**b**) PEN in the Vis-fUV spectral region

spectra are dominated by Fabry–Perot oscillations due to the multiple reflections of light at the film interfaces [9, 10, 24].

Among the more intense characteristic vibration bands in the FTIRSE spectra of PET, (Fig. 7.6a), we observe the vibration modes at ~940 and ~971 cm^{-1} (trans) that could be attributed to the C–O stretching mode, the aromatic CH_2 stretching mode at ~1,125 cm^{-1}, the ester mode at ~1,255 cm^{-1}, the in-plane deformation of the C–H bond of the para-substituted benzene rings at ~1,025 and ~1,410 cm^{-1} and furthermore, the characteristic vibration band at 1,720 cm^{-1} corresponding to the stretching vibration of the carbonyl C=O groups. The band at 1,342 cm^{-1} is attributed to the wagging mode of the ethylene glycol CH_2 groups of the *trans* conformations. Also, we observe at 1,470 cm^{-1} the characteristic peak corresponding to the CH_2 bending mode, whereas the C–H in plane deformation mode appears at ~1,505 cm^{-1} [9, 10, 24].

Due to the existence of naphthalene ring structure in the monomer unit of PEN instead of a benzene ring structure, in PET, the FTIRSE spectra (Fig. 7.6b)

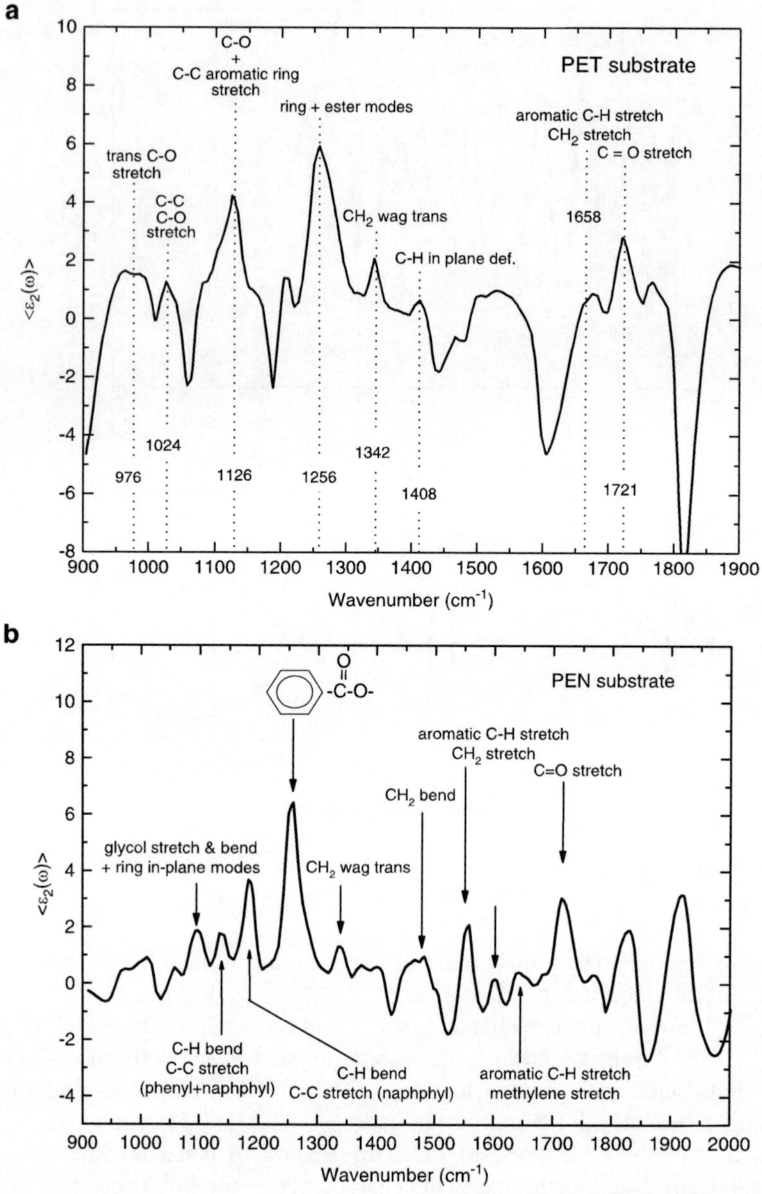

Fig. 7.6 Measured $\langle \varepsilon(\omega) \rangle$ of (**a**) PET and (**b**) PEN flexible substrates in the IR spectral region

shows a similar IR response, however, with some additional vibration bands. These include the bands at 1,098 cm^{-1} that has been attributed to the stretching and bending modes of ethylene glycol attached to the aromatic structures of the PEN monomer units. Moreover, the characteristic band at 1,184 cm^{-1} corresponds to the C–C stretching modes of the naphthalene group. The complex bands at 1,335 and 1,374 cm^{-1} reveal the bending mode of the ethylene glycol CH_2 group in the gauche and trans conformations, respectively. The C=C stretching modes of the aromatic (naphthalene) ring structures of PEN can be observed at $\sim 1,635\,cm^{-1}$. Moreover, the stretching vibration of the carbonyl C=O group appears in lower energy in case of PEN (1,713 cm^{-1}) than in PET (1,720 cm^{-1}). This could be the result of the increased conjugation due to the existence of naphthalene (PEN) instead of benzene (PET) rings structures, which shifts the maximum absorbance energy to lower values [9, 10, 24].

Finally, the investigation of the optical properties of PET and PEN substrates in the spectral region, from the IR to the NIR-Vis-fUV, allows the calculation of the bulk dielectric function $\varepsilon(\omega)$. The Fig. 7.7 shows the calculated bulk $\varepsilon(\omega)$ of PET and PEN substrates. This is calculated at the high symmetry orientation $\theta = 0°$ (plane of incidence parallel to the MD), using the best-fit parameters obtained by the analysis of the measured $\langle\varepsilon(\omega)\rangle$ taking account the peaks I–IV (Vis-fUV) and the characteristic bands corresponding to the more intense bonding vibrations in the wavenumber region 900–1,800 cm^{-1} [9, 10, 24].

The high energy dielectric constant that corresponds to the optical transitions at energies $\omega \geq 6.5$ eV, has the values of 2.39 (PET) and 2.44 (PEN), which are

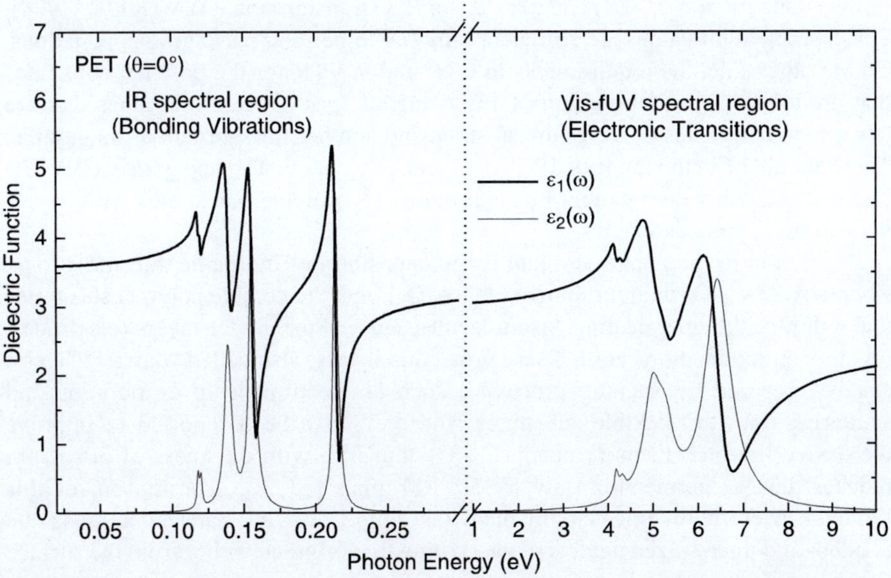

Fig. 7.7 Calculated Bulk $\varepsilon(\omega)$ of PET substrate in the energy region from the IR to the Vis-fUV

higher than unity. This indicates that more electronic transitions should be expected at higher energies. Indeed, it has been found at the literature, that both PET and PEN polymer substrates show electronic transitions at energies (around 13.3 and 15.5 eV), which are characterized by optical anisotropy. These band-to-band transitions do not affect the optical absorption measured up to 6.5 eV measurement limit, however contribute to the $\varepsilon_1(\omega)$, inducing an increase from unity [9, 10, 24].

From the above it is clear that the implementation of SE as a nanometrology tool for the investigation of the optical and electronic properties of flexible polymer films is of major importance and provides significant information on the understanding of their optical, structural, and vibrational properties. This information is required for the study of the active (e.g., small molecule and polymer organic semiconductors) and passive materials (electrodes, barrier layers) that can be deposited onto the polymeric films used as substrates for organic electronics applications.

7.4.2 *Barrier Layers for Encapsulation of Devices*

The stability of the organic semiconductors and the electrodes to corrosion due to the permeation of the atmospheric gas molecules (O_2 and H_2O) is one of the most important challenges that have to be addressed and overcome in order for FEDs to be suitable for market exploitation [1]. Also, the permeation of the atmospheric gases leads to the film delamination which results in the device failure. The commercial polymer films that are used as flexible substrates (for example) exhibit relatively high permeability values, which are measured in the range of 10^{-1}–10^{2} cm^3/m^2dbar (for O_2 transmission – OTR) and g/m^2d (for H_2O transmission – WVTR) [1,27–29]. These permeation values are sufficient only for some food packaging applications. On the other side, the requirements in OTR and WVTR for the polymer substrates that are used for the fabrication of FEDs include gas permeation values that are lower than the values of the plain substrates by another three orders of magnitude. These should be in the range of 10^{-4}–10^{-5} cm^3/m^2dbar (OTR) and g/m^2d (WVTR) [1, 27–29]. Such values cannot be achieved by the polymer substrates with any of the current r2r processes [30].

A solution to the above problem is the deposition of inorganic thin films (e.g., silicon oxide – SiO_x or aluminum oxide – AlO_x) onto the flexible polymer substrates that will block the permeating gas molecules and seal the device layers (electrodes, organic semiconductors, etc.). These protection layers (also called "barrier" layers) can be prepared by vacuum processes, such as electron beam evaporation and sputtering onto the flexible substrates, and they have been reported to improve the device lifetime. In more detail, a SiO_x thin film with thickness of 30–50 nm reduces the gas permeation only by 50–100 times [31, 32]. The reason for this limited barrier improvement attainable with a single inorganic barrier coating is due to nano- and micro-sized defects in the coating that originate either from the surface roughness of the underlying substrate or from the inorganic coating processing conditions [27].

However, the barrier response of the inorganic (e.g., SiO_x) thin films deposited onto flexible substrates depends on various factors, such as the process parameters, the deposition rate and the initial stages of the deposition and the stiffness and smoothness of the substrate [28, 33]. The nucleation and initial growth, has a major impact on the final film structure affecting the barrier response. The more uniform the nucleation on the surface, the more likely it is that the SiO_x film will be homogeneous in structure and the less likely that defects will be introduced during growth. Therefore, by increasing the uniformity and the density of nucleation on a polymer surface may improve the barrier performance. Therefore, it is very important to understand the nucleation and growth of the oxide on polymer substrates, and in particular, the effect of the polymer surface characteristics on the processes.

This information can be obtained by the optical investigation of the inorganic film growth by the use of in situ and real-time SE in the Vis-fUV spectral region [11, 12]. The analysis of the real-time SE measurements can provide information about the optical properties, the deposition rate, the stoichiometry, and the growth mechanism and how it is affected by surface of the substrate.

Figures 7.8 and 7.9 show the evolution of the imaginary part $\langle \varepsilon_2(\omega) \rangle$ of the measured pseudodielectric function $\langle \varepsilon(\omega) \rangle$ as measured during the deposition of the SiO_x thin film by electron beam evaporation onto PET and PEN flexible polymer substrates. The PET and PEN substrates are the commercially available Melinex 401 and Teonex substrates with thickness of 50 μm and 125 μm, respectively. The working pressure was adjusted to Pw $= 10^{-6}$ Torr, whereas the accelerating voltage and the beam current were fixed at 7 kV and 100 mA, respectively [28]. The

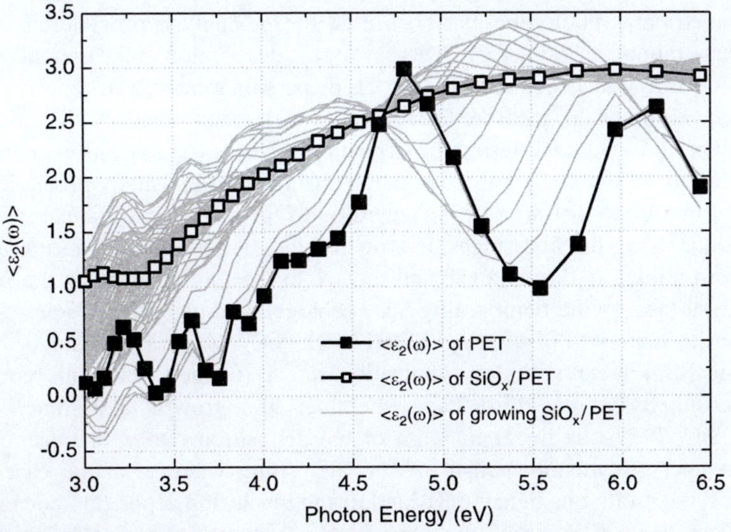

Fig. 7.8 Real-time spectrum of the $\langle \varepsilon_2(\omega) \rangle$ spectra recorded during the deposition of SiO_x film onto PET substrate

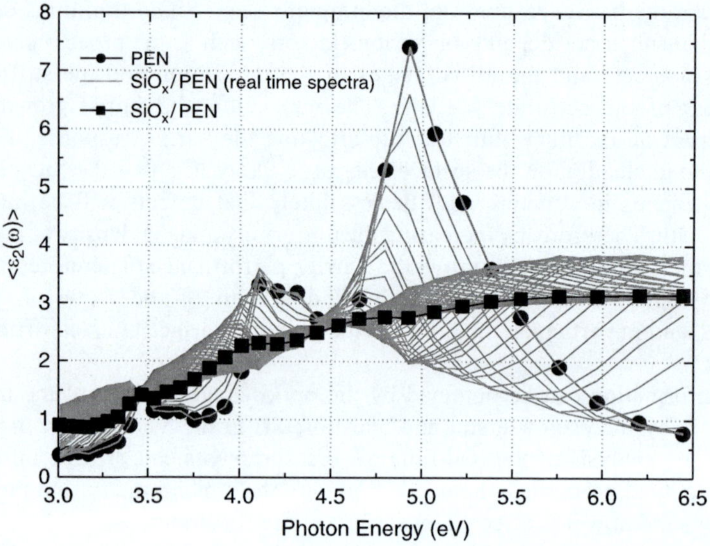

Fig. 7.9 Real-time spectrum of the $\langle \varepsilon_2(\omega) \rangle$ spectra recorded during the deposition of SiO_x film onto PEN substrate

optical properties of the growing SiO_x thin films were measured by an ultra-fast Multi-Wavelength Phase Modulated SE unit that is adapted on the UHV chamber with an angle of 70° [28].

The parameterization of the optical and electronic properties of the SiO_x films has been performed by the modeling of the measured $\langle \varepsilon(\omega) \rangle$ during the film deposition. The geometrical structure that has been used for the analysis procedure includes a three-phase model: air/SiO_x/PET, whereas for the description of the optical response of the SiO_x film it has been used for the TL dispersion model [21].

The modeling of the measured real-time $\langle \varepsilon(\omega) \rangle$ spectra leads to the calculation of the SiOx film thickness during the deposition process. As it is has been reported, the evolution of the SiO_x film thickness during the deposition process can be separated into three distinct stages, according to the growth mechanisms that take place [28]. During the first stages of growth it can be observed an abrupt increase of the film thickness that is attributed mostly to the modification of the polymer substrate surface by the bombarding SiO_x (charged and neutral) particles, forming a composite material constituting a PET with SiO_x. At the next stage, a nano-crystallite SiO_x layer with low deposition rate is formed, onto this composite material. Finally, at the third stage, the final film growth takes place. In the case of SiO_x/PEN, the linear increase of the deposition rate with time reveals a layer-by-layer growth mechanism of the SiO_x film, whereas in the case of the SiO_x/PET, the oscillating behavior (island formation during deposition rate increase, and coalescence stages during the deposition rate decrease) of the deposition rate

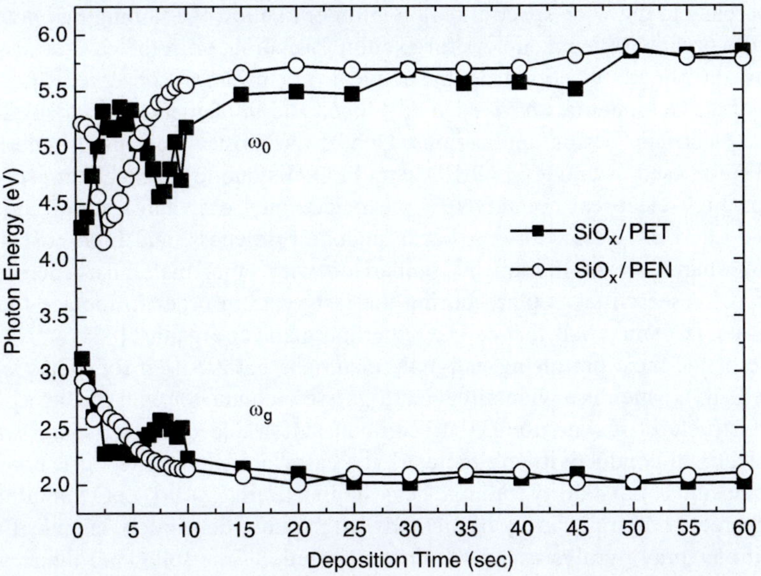

Fig. 7.10 Fundamental gap ω_g and Penn gap ω_0 as determined by the real-time analysis of the measured $\langle \varepsilon(\omega) \rangle$ spectra during the deposition of SiO_x onto PET and PEN substrates

indicates that the SiO_x film growth is dominated by an island growth mechanism [28, 34].

Figure 7.10 shows the calculated fundamental gap ω_g and energy position of maximum in $\varepsilon_2(\omega)$ or Penn gap ω_0 during the deposition of SiO_x/PET and SiO_x/PEN. In the case of SiO_x/PET, the calculated ω_g is relatively stable after the first 25 s of deposition.

In the case of SiOx/PEN the ω_g values are found to be higher at the nucleation stage, whereas in the stage of homogeneous growth both the ω_g and ω_0 values are stable. In addition, during the initial stages of SiO_x/PEN, it has been reported that ω_0 decreases during the nucleation stage ($t < 2.5$ s) and increase during the stage of coalescence ($2.5\,\text{s} < t < 9\,\text{s}$) [28, 34].

7.4.3 Transparent Electrodes (Inorganic, Organic)

An important part of the organic electronic device is the electrodes which have a major role for the transport of charge carriers, providing to the device its functionality. Although for the conventional microelectronic devices, electrodes are consisted only of metals, such as Al, Ca, etc., in the case of flexible organic electronics there are more and strict requirements and properties that the electrode materials should fulfil [1, 2, 35, 36]. For example, these materials should be characterized by high transparency in the wide spectral region in order to allow the impinging photons to reach the organic semiconductors for exciton formation, or in the case of displays, to allow the photon injection from the active layers toward the observer [2].

Several TCO materials have been developed and implemented as electrodes for organic electronic device applications. One of the most widely used materials is ITO. ITO is used as anode in OLEDs and in OPVs due to its excellent properties such as high electrical conductivity, piezoelectricity, easy fabrication, and non-toxicity [2, 3]. However, its drawbacks include brittleness, and high cost due to the low abundance of Indium in comparison with other materials. Therefore, a systematic research takes place during the last years in order to replace the ITO with other TCO materials that can be either inorganic or organic [1–3].

One of the most promising inorganic materials that are used for TCOs is Zinc Oxide (ZnO), which is a wide direct band-gap semiconductor with the hexagonical crystal structure of wurtzite [2]. It combines desirable characteristics, such as high electrical conductivity, piezoelectricity, easy fabrication, low cost, non-toxic character, and compatibility for large-scale applications [37–39]. ZnO thin films can be prepared by a large variety of methods with sputter deposition, chemical vapor deposition, spray pyrolysis, and vacuum evaporation. The following figure shows the unit cell of ZnO [40] (Fig. 7.11).

Several efforts have been performed in order to understand the growth mechanisms of ZnO films onto rigid and flexible substrates, their functionality and combination with organic–inorganic materials as well as the effect of the deposition parameters in their optical, structural, and electronic response [1, 2, 37, 40–47]. In situ and real-time SE can be applied to investigate the optical properties of ZnO thin films during their deposition onto rigid and polymer substrates. The optical measurements were performed using an ultra-fast multi-wavelength phase modulated SE unit, adapted onto the UHV chamber at an angle of 70°.

Figure 7.12 shows the evolution with time of the $\langle \varepsilon_2(\omega) \rangle$ in the 3–6.5 eV spectral region during the deposition of ZnO film onto a PEN substrate by Pulsed DC Magnetron Sputtering (pulse frequency at 100 kHz). The $\langle \varepsilon_2(\omega) \rangle$ were recorded in real-time during the deposition of ZnO films onto the PEN substrates with thickness of 125 μm and surface roughness of 1.75 nm. The deposition time is 1,200 s whereas the final film thickness is 72 nm [40].

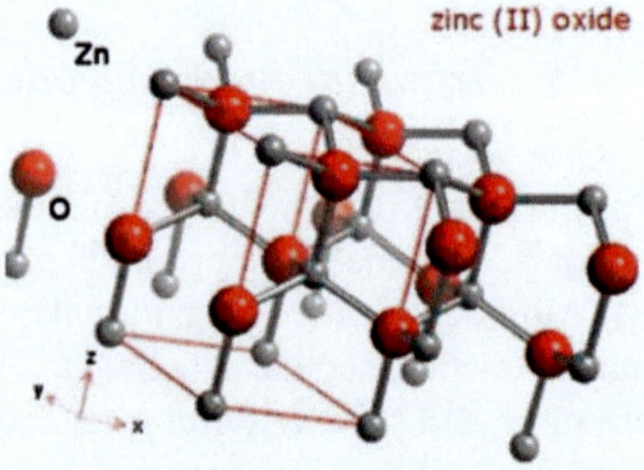

Fig. 7.11 Unit cell of ZnO

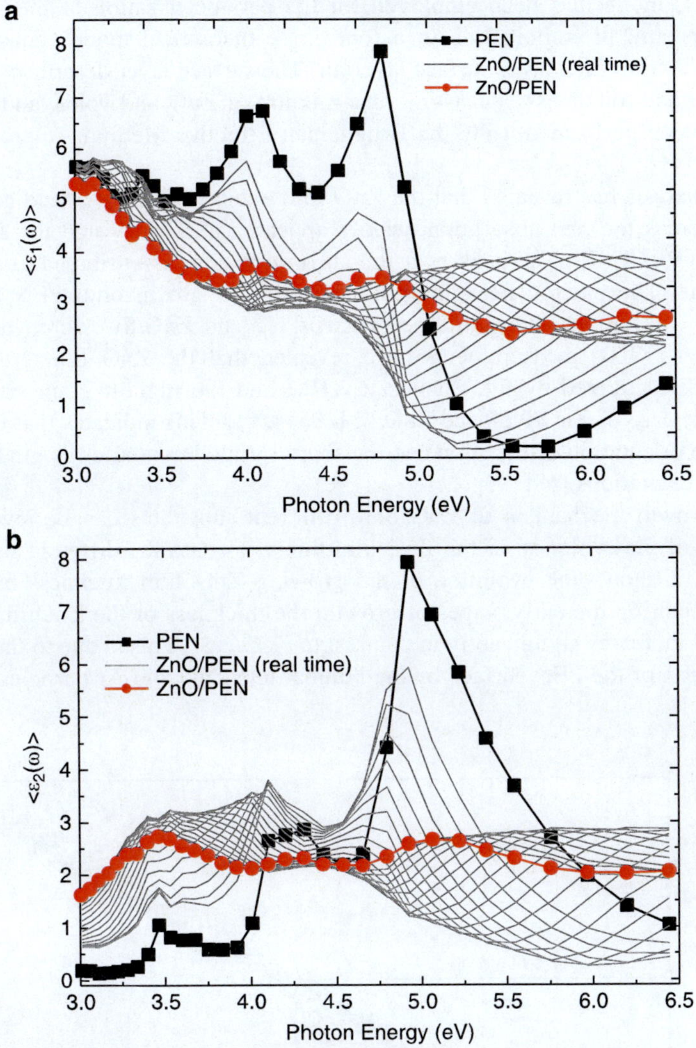

Fig. 7.12 Real-time measured $\langle \varepsilon_2(\omega) \rangle$ of ZnO film grown onto a PEN flexible substrate by pulsed DC magnetron sputtering. The spectra were measured before, during, and after the deposition process [40, 48]

From Fig. 7.12 we observe an increase of the intensity of the exciton absorption peak (ω_{01}) of ZnO film during the deposition is observed. At first, the growing crystallites are small in size and contain many defects, obstructing the excitations (the exciton Bohr radius is calculated at 1.8 nm). As the film thickness and grain size increase, there is a quality improvement of grains with lower amount of defects, favoring the appearance of the optical absorptions at the $\langle \varepsilon_2(\omega) \rangle$ spectra.

The TL model has been employed for the parameterization of the measured $\langle\varepsilon(\omega)\rangle$ spectra, in combination to a four-phase theoretical model, consisting of substrate/bottom ZnO layer/surface layer/air. The surface layer describes the effect of the surface roughness, consisting of a mixture of ZnO and voids, and its analysis has been performed using the Bruggeman Effective Medium Approximation (BEMA) [11].

The analysis has revealed that the ZnO film exhibit its energy band gap above 3 eV, whereas the first absorption peak is appeared at 3.17 eV and the second at ~6.4 eV [40, 44, 48]. The peak near the absorption edge is attributed to excitonic excitations. The exciton peak intensity of the ZnO grown onto PEN substrate is found to be higher than the case of ZnO/PET and ZnO/Si. Moreover, X-Ray Diffraction (XRD) measurements have revealed that the ZnO film grown onto PEN is characterized by the smallest FWHM and the highest grain size values than in the case of ZnO/PET and ZnO/Si [40, 44, 48]. This indicates that the better crystallite orientation distribution and the shorter grain boundaries length favor the excitonic excitations [48].

The growth mechanism of ZnO onto different substrates can be revealed by the study of the evolution of the ZnO film thickness onto the different substrates. Figure 7.13 shows the evolution of the growing ZnO film thickness onto PEN and c-Si. During the early stages of growth, the thickness of the ZnO film grown onto PEN increases significantly in contrast to ZnO/c-Si. This is due to the surface modification of the PEN surface by the bombarding ZnO and Ar particles (neutral

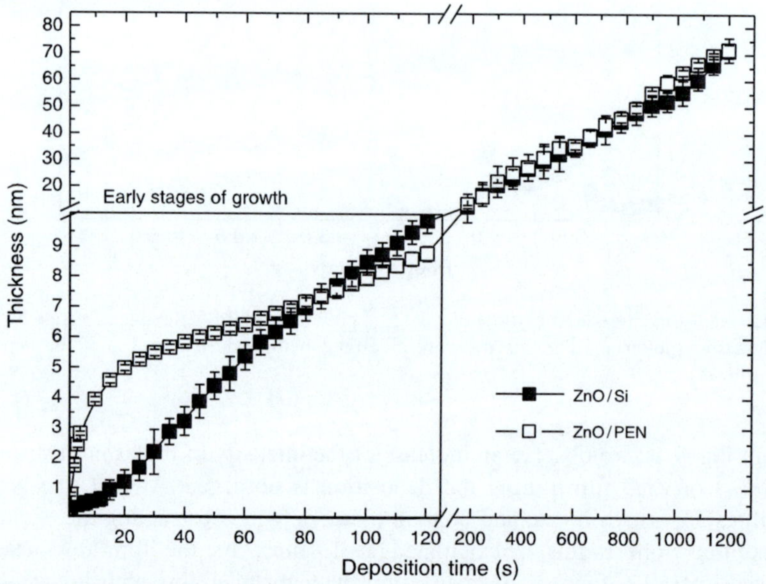

Fig. 7.13 Evolution of the film thickness of ZnO films deposited onto c-Si and PEN substrates by pulsed dc sputtering, as determined by SE [44, 48]

and charged) and to the formation of an interface layer consisting of modified PEN and ZnO. This is confirmed by High Resolution Transmission Electron Microscopy (HRTEM) measurements that revealed a composite nanocrystalline overlayer on top of the polymer surface [48]. However, this effect is not pronounced on the c-Si substrate on which the ZnO film is grown following a layer-by-layer mechanism.

After this stage, the deposition rate of ZnO/PEN increases to 0.04 nm/s, indicating a more effective deposition onto the modified surface, whereas at the later stage of growth the ZnO film growth is dominated by a layer-by-layer mechanism (deposition rates: 0.04 and 0.07 nm/s for PEN and Si, respectively) until the end of the deposition process (1,200 s) [48].

Another material that is widely used as an anode buffer layer in FEDs is poly(3, 4-ethylenedioxythiophene) poly(styrenesulfonate) (PEDOT/PSS) [1–3,49,50]. This is a very promising transparent conductive polymer that is expected to replace the inorganic, brittle, and expensive ITO as well as other TCO materials. PEDOT:PSS is characterized by high conductivity and it is used as hole-injecting material in applications such as sensors, antistatic coatings, and solar cells. [1, 3, 49]. PEDOT:PSS consists of a conducting part PEDOT, which is a low molecular weight conjugated polymer, insoluble and thus difficult to process and an insulating polymer (PSS), which is a high molecular weight polymer that gives the desirable flexibility (see Fig. 7.14) [3, 6, 51]. PSS also increases the solubility in water, making the whole system easy to process. The oligomer PEDOT segments are electrostatically attached on the PSS polymer chains [1, 2, 6, 35, 51].

The optical properties of the PEDOT:PSS films have been measured by SE in the Vis-fUV spectral region. In order to extract quantitative information from the measured $\langle\varepsilon(\omega)\rangle$, this has been analyzed by the use of three phase geometrical model, which consists of air (ambient), the PEDOT:PSS layer (with thickness d) on top of the PET (bulk) substrate (air/PEDOT:PSS/PET substrate). For the ambient medium it has been used air in which $\varepsilon_1(\omega) = 1$ and $\varepsilon_2(\omega) = 0$ for all energy values ω.

The optical properties and the thickness of the PEDOT:PSS layer have been modeled by the use of the Tauc–Lorentz (TL) dispersion model. As it has been discussed in detail elsewhere in this model the imaginary part $\varepsilon_2(\omega)$ of the dielectric function is obtained multiplying the equation of the Lorentz oscillator by the equation of the Tauc joint density of states, and the real part $\varepsilon_1(\omega)$ is determined

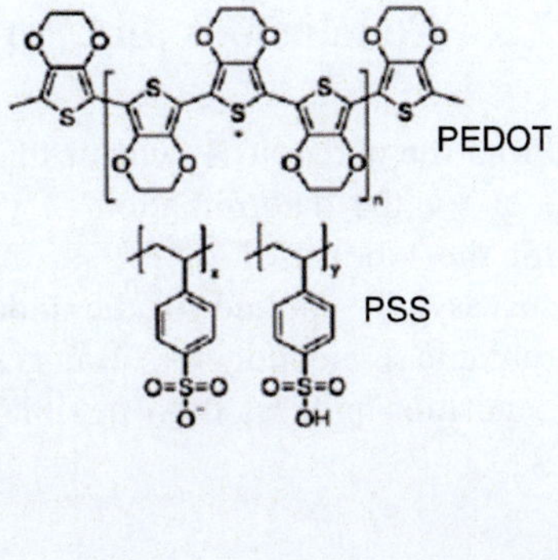

Fig. 7.14 Structure of the PEDOT:PSS

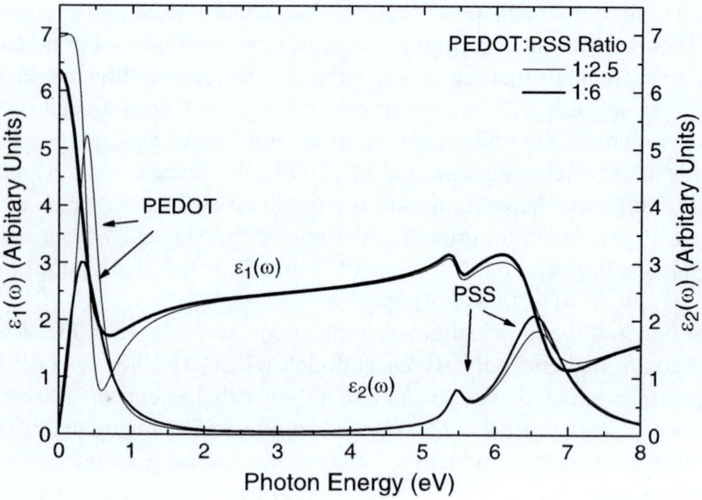

Fig. 7.15 Calculated bulk dielectric function $\varepsilon(\omega)$ of the PEDOT:PSS films with various PSS weight ratios

by the imaginary part $\varepsilon_2(\omega)$ by the Kramers–Kronig integration [14, 17]. For the parameterization of the optical properties and thickness of the PEDOT:PSS films, we took into consideration only the characteristic optical absorptions of the π–π^* transitions of the benzene rings of PSS that appear at 5.3 eV and 6.33 eV, respectively [35,36,52,53]. This modeling procedure provides in detail the thickness values of the PEDOT:PSS films. Figure 7.15 shows the determined bulk dielectric function $\varepsilon(\omega)$ of the PEDOT:PSS film in the IR–Vis–fUV spectral region with various PSS weight ratios [35]. This shows three absorptions in energies 0.47, 5.37, and 6.38 eV. The absorbance peaks at high energies can be attributed to the π–π^* transitions of the benzene rings of PSS and the absorption in low energy attributed to the PEDOT [6, 36, 54]. The results show that the increase of PSS ratio in PEDOT:PSS solution leads to the reduction of ω_{01} optical absorption attributed to PEDOT (decrease of oscillator's amplitude) and to a slight increase of ω_{02} and ω_{03} absorptions attributed to the π–π^* transitions of the benzene rings of PSS [6,36,54].

7.5 Conclusions and Perspectives

From the above it is evident that SE has the potentiality to be used as a standard tool for the determination of the optical properties of state-of-the-art materials for the fabrication of FEDs. In more specific, in situ and real-time SE can be successfully applied for the understanding of the growth mechanisms of materials (organic semiconductors, transparent electrodes, barrier materials, etc.) during their deposition process onto flexible polymeric substrates. The optical properties can be measured in an extended spectral region from the IR to the Vis-fUV spectral region in order to stimulate different light-matter mechanisms (bonding vibrations, interband transitions), which will provide significant insights on their properties.

Also, this powerful technique can be implemented to production lines for FED applications as a quality control tool. Its robustness and high flexibility for adaptation gives numerous capabilities for the improvement of the produced devices as well as for the optimization of the production process in terms of cost and waste.

References

1. S. Logothetidis, A. Laskarakis, Eur. Phys. J. Appl. Phys. **12502** 12502 (2009)
2. S. Logothetidis, A. Laskarakis, Thin Solid Films **518** 1245–1249 (2009)
3. N. Koch, ChemPhysChem. **8** 1438–1455 (2007)
4. S. Bertho, W.D. Oosterbaan, V. Vrindts, T.J. Cleij, L. Lutsen, J. Manca, D. Vanderzande, Org. Electron. **10** 1248–1251 (2009)
5. O. Gordan, M. Friedrich, D. Zahn, Org. Electron. **5** 291–297 (2004)
6. L.A. Pettersson, S. Ghosh, O. Inganas, Org. Electron. **3** 143–148 (2002)
7. S.P. Speakman, G.G. Rozenberg, K.J. Clay, W.I. Milne, A. Ille, I.A. Gardner, E. Bresler, J.H. Steinke, Org. Electron. **2** 65–73 (2001)
8. M.I. Alonso, M. Garriga, N. Karl, J.O. Oss, F. Schreiber, Org. Electron. **3** 23–31 (2002)
9. A. Laskarakis, S. Logothetidis, Appl. Surf. Sci. **253** 52–56 (2006)
10. A. Laskarakis, S. Logothetidis, J. Appl. Phys. **101** 1–9 (2007)
11. R. Azzam, N. Bashara, *Ellipsometry and Polarized Light* (North Holland, Amsterdam, 1977)
12. G.E. Irene, H.G. Tompkins (Eds.) *Handbook of Ellipsometry* (William Andrew, Norwich, 2005)
13. A. Laskarakis, S. Logothetidis, C. Charitidis, M. Gioti, Y. Panayiotatos, M. Handrea, W. Kautek, Diam. Relat. Mater. **10** 1179–1184 (2001)
14. G.E. Jellison, Thin Solid Films **313–314** 33–39 (1998)
15. M. Losurdo, M.M. Giangregorio, P. Capezzuto, G. Bruno, F. Babudri, D. Colangiuli, G.M. Farinola, F. Naso, Synth. Met. **138** 49–53 (2003)
16. M. Schubert, Thin Solid Films **313–314** 323–332 (1998)
17. H.G. Tompkins, Thin Solid Films **455–456** 772–778 (2004)
18. G.E. Jellison, Thin Solid Films **290–291** 40–45 (1996)
19. A. Laskarakis, S. Logothetidis, M. Gioti, Phys. Rev. **64** 1–15 (2001)
20. S. Logothetidis, A. Laskarakis, A. Gika, P. Patsalas, Surf. Coat. Technol. **152** 204–208 (2002)
21. G.E. Jellison, F.A. Modine, Appl. Phys. Lett. **69** 371–373 (1996)
22. G.E. Jellison, F.A. Modine, P. Doshi, A. Rohatgi, Thin Solid Films **313–314** 193–197 (1998)
23. A. Laskarakis, D. Georgiou, S. Logothetidis, Phys. Stat. Sol. A 1–7(2010)
24. A. Laskarakis, S. Logothetidis, J. Appl. Phys. **99** 066101–1–066101–3 (2006)
25. M. Gioti, A. Laskarakis, S. Logothetidis, Thin Solid Films **456** 283–287 (2004)
26. A. Laskarakis, D. Georgiou, S. Logothetidis, Mater. Sci. Eng. B **166** 7–13 (2010)
27. D. Georgiou, A. Laskarakis, S. Logothetidis, U. Weber, M. Schmidt, K. Noller, Appl. Surf. Sci. **255** 8023–8029 (2009)
28. D. Georgiou, S. Logothetidis, C. Koidis, A. Laskarakis, Phys. Stat. Sol. (C) **4** 1–4 (2008)
29. S. Logothetidis, A. Laskarakis, D. Georgiou, S. Amberg-Schwab, U. Weber, Phys. Stat. Sol. (C) 1–5 (2008)
30. K. Haas, S. Amberg-Schwab, K. Rose, Thin Solid Films **351** 198–203 (1999)
31. C. Charton, N. Schiller, M. Fahland, A. Holla, A. Wedel, K. Noller, Thin Solid Films **502** 99–103 (2006)

32. U. Weber, A. Burger, S. Nique, R. Xalter, Monatshefte Fur Chemie (Chem. Month.) **666** 657–666 (2006)
33. M.S. Hedenqvist, K.S. Johansson, Surf. Coat. Technol. **172** 7–12 (2003)
34. D. Georgiou, A. Laskarakis, C. Koidis, N. Goktsis, S. Logothetidis, Phys. Stat. Sol. (C) **3391** 3387–3391 (2008)
35. C. Gravalidis, A. Laskarakis, S. Logothetidis, Eur. Phys. J. Appl. Phys. **46** 12505 (2009)
36. A.M. Nardes, M. Kemerink, R.A. Janssen, Phys. Rev. B 085208–1–085208–7 (2007)
37. S. Logothetidis, A. Laskarakis, S. Kassavetis, S. Lousinian, C. Gravalidis, G. Kiriakidis, Thin Solid Films **516** 1345–1349 (2008)
38. S. Tsai, Y. Lu, J. Lu, M. Hon, Surf. Coat. Technol. **200** 3241–3244 (2006)
39. W. Jeong, S. Kim, G. Park, Thin Solid Films **506–507** 180–183 (2006)
40. C. Koidis, S. Logothetidis, D. Georgiou, A. Laskarakis, S. Lousinian, I. Tsiaoussi, N. Frangis, Phys. Stat. Sol. (A) **205** 1988–1992 (2008)
41. A. Pimentel, E. Fortunato, A. Goncalves, A. Marques, H. Aguas, L. Pereira, I. Ferreira, R. Martins, Thin Solid Films **487** 212–215 (2005)
42. E. Fortunato, P. Barquinha, A. Pimentel, A. Goncalves, A. Marques, L. Pereira, R. Martins, Thin Solid Films **487** 205–211 (2005)
43. C. Koidis, S. Logothetidis, S. Kassavetis, A. Laskarakis, N.A. Hastas, O. Valassiades, Phys. Stat. Sol. A 1–5 (2010)
44. C. Koidis, S. Logothetidis, A. Laskarakis, I. Tsiaoussis, N. Frangis, Micron **40** 130–134 (2009)
45. J. Hupkes, B. Rech, S. Calnan, O. Kluth, U. Zastrow, H. Siekmann, M. Wuttig, Thin Solid Films **502** 286–291 (2006)
46. M. Suchea, S. Christoulakis, K. Moschovis, N. Katsarakis, G. Kiriakidis, Thin Solid Films **515** 551–554 (2006)
47. E. Fortunato, P. Barquinha, A. Pimentel, A. Goncalves, A. Marques, L. Pereira, R. Martins, Thin Solid Films **487** 205–211 (2005)
48. C. Koidis, S. Logothetidis, D. Georgiou, A. Laskarakis, Phys. Stat. Sol. (C) **5**(5) 1366–1369 (2008)
49. S. Jonsson, J. Birgerson, X. Crispin, G. Greczynski, W. Osikowicz, W.R. Salaneck, M. Fahlman, Synth. Met. **139** 1–10 (2003)
50. Z. Tang, S.T. Donohoe, J.M. Robinson, P.A. Chiarelli, H. Wang, Polymer **46** 9043–9052 (2005)
51. T.P. Nguyen, P.L. Rendu, P.D. Long, S.A. Vos, Surf. Coat. Technol. **180–181** 646–649 (2004)
52. V. Shrotriya, J. Ouyang, R.J. Tseng, G. Li, Y. Yang, Chem. Phys. Lett. **411** 138–143 (2005)
53. K. Yim, R. Friend, J. Kim, J. Chem. Phys. **124** 184706 (2006)
54. T. Johansson, Synth. Met. **101** 198–199 (1999)

Chapter 8
Introduction to Organic Vapor Phase Deposition (OVPD®) Technology for Organic (Opto-)electronics

Dietmar Keiper, Nico Meyer, and Michael Heuken

Abstract In this chapter, the organic vapor phase deposition (OVPD®) technology combined with the Close Coupled Showerhead (CCS) technology for the fabrication of sophisticated opto-electronic organic devices based on open literature will be shortly reviewed. Typically, organic (opto-)electronic devices are fabricated by vacuum thermal evaporation (VTE), which is in contrast with the OVPD® technology. The deposition of single organic films, the morphology control by OVPD and the proposed benefits of mixing organic materials, and applying non-sharp interfaces for the overall organic light emitting diode (OLED) performance will be discussed.

8.1 Introduction

The first two-layered organic light emitting diode (OLED) consisting of α-NPD and Alq_3 was reported by Kodak in 1987 [1]. Since that time organic (opto-) electronic devices like Thin Film Transistors (TFTs), solar cells, photo detectors and especially OLEDs for display and solid-state lighting applications have been intensively investigated [2–6]. OLEDs feature beneficial intrinsic properties such as high brightness, low energy consumption, and wide viewing angle characteristics realized in thin devices on glass or on even flexible substrates. This makes them attractive for display application, where they are nowadays applied in cell phones or MP3 players. The next field of commercialization is solid-state lighting [7,8]. Here, OLEDs reveal additional benefits as area light source with flexible form factors in different colors ranging from red to blue or color mixtures like white light. Latest works on white emitting OLEDs significantly extended the external quantum

M. Heuken (✉) · D. Keiper · N. Meyer
AIXTRON AG, Kaiserstr. 98, 52134 Herzogenrath, Germany
e-mail: D.Keiper@aixtron.com

S. Logothetidis (ed.), *Nanostructured Materials and Their Applications*,
NanoScience and Technology, DOI 10.1007/978-3-642-22227-6_8,

efficiency (EQE) and efficacies of 102 lm/W (CIE: 0.41; 0.46) [9, 10] and even 124 lm/W (CIE: 0.45; 0.47) were reported using sophisticated organic materials, multilayer OLED stack designs together with improved light outcoupling (ILO) structures [11]. These high efficacies could only be achieved by a sophisticated, ILO approach. Despite the fact that the CIE coordinates are not on the Planckian locus these results prove the potential of OLEDs for solid-state lighting. But beside the efficacy also other, maybe interacting, aspects like lifetime, color, internal quantum efficiency, uniformity, and reproducibility need to be addressed. This needs further improvements on the applied organic materials and on the overall OLED design and fabrication.

The above described record OLED stack applied a layer structure with sharp interfaces and mixtures of in maximum two organic materials [11]. However, other OLED structures were suggested like using three organic materials in one layer and different improvements on the overall OLED were observed [12–14]. One group attributed the improvement of the OLED stack to the fact that the hole transport in the emissive layer (EML) was improved by shifting it from the dopant to the co-dopant. The latter has a lower HOMO than the dopant [9, 13]. By the latter approach a quasi ideal host with perfect HOMO and LUMO levels was created by the combination of a Host, for the electron transport, and an additional second dopant for the hole transport.

Beside the use of more than two different organic molecules in one layer, also a continuously graded transition between the different layers could improve the OLED performance and lifetime, as reported by Kido [15]. Thus, mixing of more than two organic materials and none sharp interfaces offer options to improve the overall OLED performance. Obviously, also other (opto-)electronic organic devices like OTFTs or organic photovoltaic (OPV) could benefit from such approaches.

Typically, these organic devices are fabricated by vacuum thermal evaporation (VTE) technology at chamber pressures of approximately 10^{-6} mbar where the organic material is heated in a source, e.g., crucible, and the organic vapor is deposited on the opposite positioned substrate. Due to the geometrical arrangement of evaporation source and substrate in the same chamber the deposition of material mixtures or concentration gradients with high uniformity on the substrate is limited. In contrast, using the organic vapor phase deposition (OVPD®) technology organic films with material mixtures or even complex concentration variations can be deposited with high homogeneity on the substrate. A detailed comparison of the OVPD® technology with the VTE technology is given elsewhere [16, 17].

In the following, we will shortly review the OVPD® technology combined with the Close Coupled Showerhead (CCS) technology for the fabrication of sophisticated opto-electronic organic devices based on open literature. The deposition of single organic films, the morphology control by OVPD and the proposed benefits of mixing organic materials and applying non-sharp interfaces for the overall OLED performance will be discussed.

8.2 OVPD® Basics and Industrial Concept

Forrest introduced the OVPD® principle to overcome many limitations of the standard VTE [18, 19]. Due to the gas phase transport principle of OVPD® the arrangement of evaporation and condensation on a substrate is decoupled as opposed to VTE. Thus, in OVPD the evaporation area, also called source area, and deposition area can independently be optimized [16]. A schematic of *AIXTRON's* Gen 2 OVPD® equipment is shown in Fig. 8.1. The organic source materials are placed in physically separated source containers arranged in separate furnaces remote from the deposition chamber. In this example four furnaces with three source containers each are shown; however, the amount and arrangement of source containers is flexible. The temperature of each furnace, thus of the organic materials, is controlled. As a consequence a defined vapor pressure of organic material is present in the source container, which is an enclosed volume. Individual pneumatic high-temperature valves switch the source flows transporting the organic material by the carrier gas, e.g., nitrogen, towards the hot wall deposition chamber. The amount of organic material transported to the deposition area (q_{Material}), given in mol/min, is described by the formula below and depends on the vapor pressure of the organic material ($P_{\text{Vap}} = P_{\text{vap}}(T)$), the total pressure in the source container (P_{Source}), and source flow of nitrogen carrier gas through the source (q_{Source}). Here V_{mol} is the molar volume.

$$q_m = \frac{q_s}{V_{\text{mol}}} \times \frac{p_{\text{Vap}}}{p_s - p_{\text{Vap}}}$$

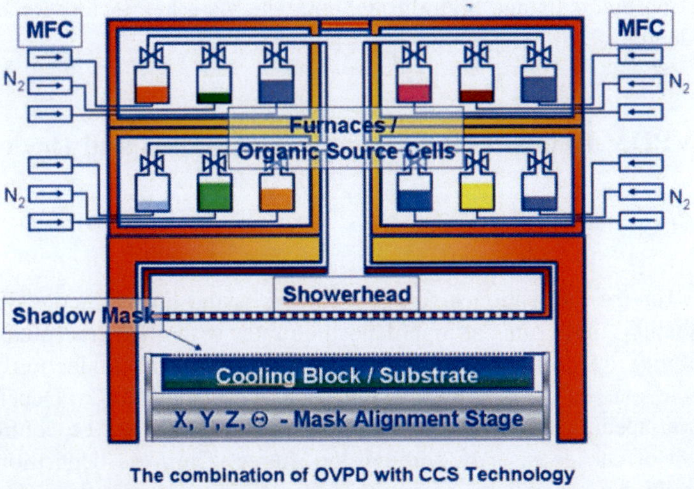

Fig. 8.1 Close Coupled Shower Head Technology for industrial scale production of organic devices by OVPD®

By altering the source flow q_{Source} using mass flow controllers (MFCs) and keeping the source temperature constant, thus P_{vap} is constant, enables a linear control of the organic material $q_{Material}$ transported to the deposition chamber [16, 18].

The individual switching of the high-temperature source valves (HT-Valves) enables a rapid on/off control of the respective deposition offering high precision control of layer interfaces as well as minimization of material waste. In addition, due to the heated runlines any unintentional condensation of organic material is avoided. The organic molecules are homogeneously mixed in the gas phase prior to be introduced uniformly through the heated showerhead across the entire substrate leading to a homogeneous condensation of the desired organic film on the substrate. The latter is placed on the cooling block and can be equipped with an alignment stage ensuring the required substrate to mask alignment. The advantages of the OVPD® technology can be fully exploited in combination with the CCS technology. The latter ensures a uniform deposition and condensation of the organic molecules over the substrate without any substrate rotation or related methods. Due to the heated lines and the controlled gas phase transport no unintended deposition of organic molecules occurs in the hot wall OVPD® chamber. Instead most of the material is deposited on the cooled substrate resulting in a high material utilization [16, 17].

This approach combining the OVPD® technology with the CCS technology enables the fabrication of a full OLED stack in a single OVPD® chamber and is not limited for scaling to any size and can be realized in a vertical or horizontal arrangement. Furthermore, this OVPD® approach for the deposition of organic films and devices offers potential for low maintenance cycles, high material yield, high reproducibility, well-defined doping with multiple dopants, films with concentration gradients and high throughput, which are key factors for industrial mass production at low cost of ownership [16, 17].

8.3 OVPD® Deposition of Organic Thin Films and Devices

8.3.1 Single Film Deposition

The basis for the fabrication of organic devices is the homogeneous deposition of high-quality, single films at high deposition rates with good controllability and reproducibility, which is important for a high tact-time and production yield. In OVPD® the source flow is precisely controlled by standard MFCs. Depending on the material specific vapor pressure the material transport can be controlled by the amount of carrier gas flow through the source. Thus, the deposition rate of each material can be adjusted as function of the source carrier flow (q_{Source}) for individual evaporation temperatures ($P_{Vap} = P_{Vap}(\text{Temp})$) and source container pressures (P_{Source}) resulting in the expected deposition rate, as shown in Fig. 8.2 for Alq_3 deposited in a Gen1 system. With increasing source flow the Alq_3 rate increases linearly up to approximately 150 sccm followed by a sub-linear increase

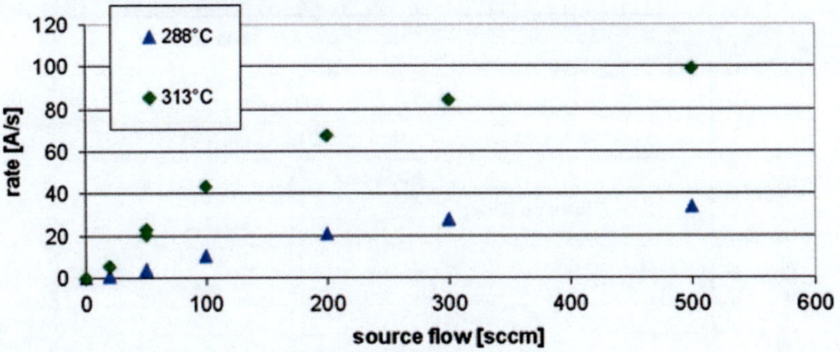

Fig. 8.2 Alq_3 rate as function of source flow with different source temperatures for a Gen1 system. The rate increases linearly with source flow up to approximately 150 sccm, followed by a sublinear increase. The rate is significantly increased with source temperature (313°C vs. 288°C) reaching 100 Å/s

towards 500 sccm. The latter is a result of the increased source container pressure P_{source} due to the higher source flow. By raising the source container temperature form 288°C to 313°C the deposition rate increases for 500 sccm up to 100 Å/s. Such high deposition rates are favorable for short tact-times. The drastic increase in rate by altered source temperature is due to the increased organic vapor pressure P_{Vap} described by the Clausius–Clapeyron equation. Worth to mention, in VTE the only option to control and increase the deposition rate is to alter the evaporation temperature. However, here a small change in source temperature has a large impact in rate. In contrast, in OVPD® the source temperature is kept at a predefined level ensuring the required range of deposition rates, but the deposition rate itself is adjusted by the nitrogen carrier gas source flow through the source container, which can be controlled in an accurate manner.

Obviously, by adding an additional source container of the same organic material at the same source temperature, the potential maximum deposition rate is doubled.

The film thickness is determined by the deposition rate and the deposition time. The latter is the time, when the source valves are open and the nitrogen carrier gas passes through the source container. The digital on/off of the HT-Valves enables the exact determination of the film thickness and sharp interfaces of subsequent organic layers. Figure 8.3 illustrates the in-situ measured Alq_3 rate for different source flows ranging from 20 sccm to 500 sccm and open (HT-Valve OPEN = 1) as well as closed source valves. With opening and closing of the source valve the in-situ measured deposition rate digitally increases and decreases, ensuring a precise control and reproducibility of the deposited film thickness. This digital on/off promotes sharp interfaces of subsequent organic layers. With increased source flow the Alq_3 deposition rate increases. Applying the same source flow the same deposition rate is achieved, obviously, as shown here for a rate of 3.9 Å/s at a source flow of 50 sccm. These results demonstrate that abrupt interfaces and reproducible thickness parameters can be realized with the control of the pneumatic source valves and precisely controlled nitrogen carrier gas flows [16, 20].

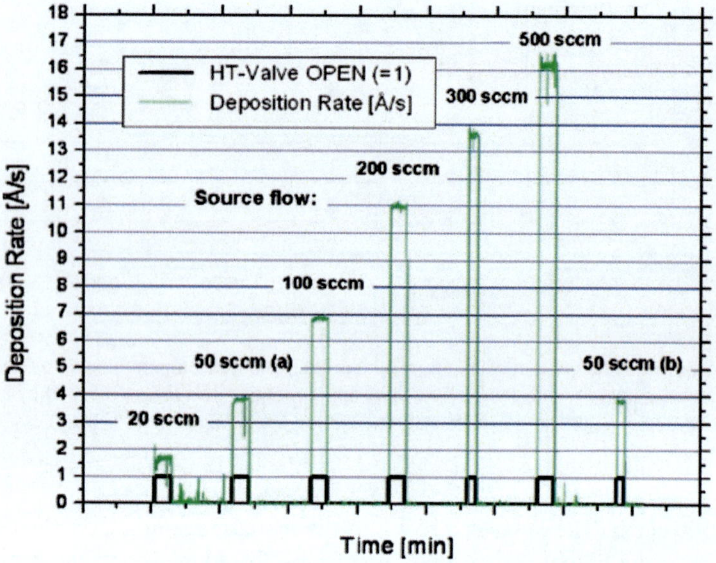

Fig. 8.3 Alq_3 deposition rate measured in-situ as function of time for different nitrogen source flows and HT-Valve switching (on/off). The Alq_3 rate shows a digital on/off and increases with source flow

In addition, not only the short-term stability but also the long-term stability of the source conditions and deposition rates was investigated in Gen1 and Gen2. For example for the latter at a rate of 12.28 Å/s a run-to-run deviation of $\sigma = 0.3\%$ was observed and over 500 h of continuous operation time at identical process parameters a standard deviation of only $\sigma = 1.93\%$ was measured [16].

The uniformity of the deposited organic films is crucial for a high production yield. OVPD® ensures by applying the CCS technology that the organic film is deposited with maximum uniformity with respect to thickness deviations over the substrate. For example for Gen1 systems thickness deviations below 2% and even below 1% were measured. In parallel for Gen2 systems an absolute thickness deviation of 1.7% over the diagonal with standard deviation $\sigma = 1\%$ were achieved [16].

These experimental uniformity values confirmed our predictions based on fluid dynamic simulations together with the material utilization efficiency for Gen2 of 50% or more. Based on these experiences the deposition chamber including showerhead was scaled and further developed towards Gen4 size (730 × 920 mm^2) and the thickness uniformity evaluated. Figure 8.4 shows the simulated deposition rate distribution over the Gen4 substrate. Here a deposition rate of 24 Å/s is shown as red color in the graph, 23 Å/s as light green and 22 Å/s as light blue color. The rates in between are interpolation from these colors. The entire Gen4 substrate shows a pure light green color illustrating that the deposition rate is very uniform over the entire substrate. The inset in the Graph shows the rate deviation assuming different edge exclusions. For a small edge exclusion of 3 mm a rate deviation of 0.52%

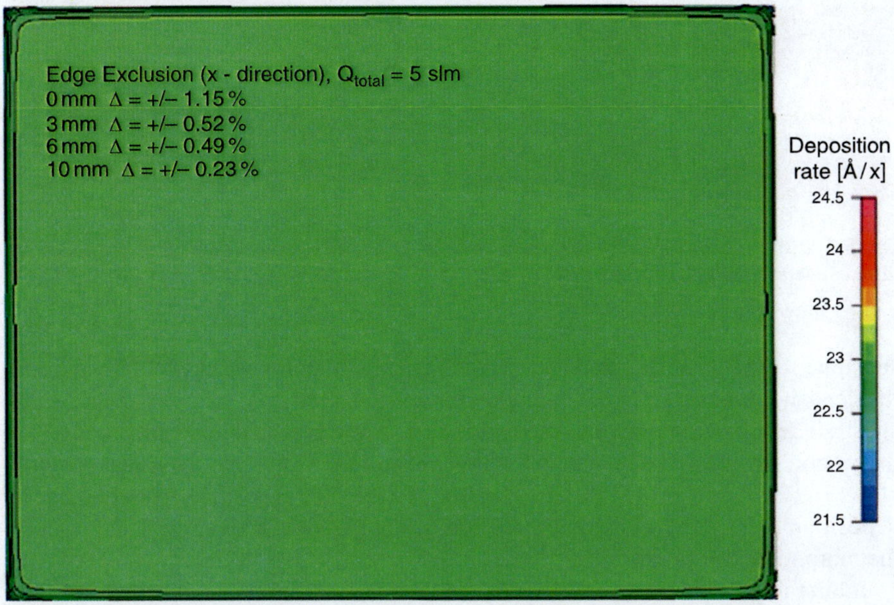

Fig. 8.4 Simulated film thickness uniformity of Alq_3 deposited on a (730 × 920) mm^2 glass substrate

was achieved. Even in case of no edge exclusion at all (0 mm) the rate deviates only by 1.15%. This feasibility study shows that high uniformities below 2% or even below 1% can be achieved proving the good scalability of the OVPD®-CCS technology, Fig. 8.4. In addition, the material utilization efficiency is expected to increase significantly from Gen2 towards Gen4 [20].

In conclusion, we have demonstrated that the combination of CCS and OVPD® enables highly reproducible deposition rates with remarkable thickness uniformities. Extending this precise process controllability of a single material to several materials offers a valuable technology for the co-evaporation of several materials for example mixed hosts or accurate doping of dyes as will be discussed in the following.

8.3.2 Organic Film Morphology

The layer morphology, thus the physical properties, of the deposited layer are affected not only by the substrate but also by the deposition process itself and the deposition parameters, like deposition rate or substrate temperature [16, 21–23]. In VTE the source (crucible) is placed opposite the substrate. To increase the deposition rate, the crucible is heated to a higher temperature leading to higher heat radiation which passively increases the substrate temperature. Thus rate and

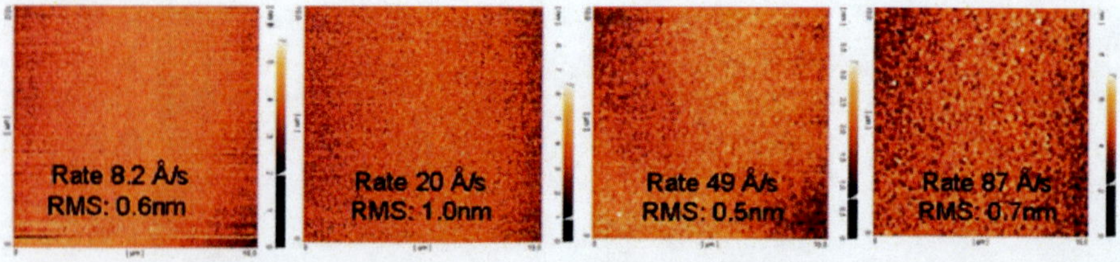

Fig. 8.5 RMS roughness of 50-nm thick α-NPD films on silicon substrates deposited with different deposition rates ranging up to 87 Å/s

substrate temperature are changed simultaneously. In contrast, in OVPD® altering the deposition rate, e.g., by changing the nitrogen source flow, has no impact on the substrate temperature due to the separation of source area and deposition area. The deposition parameters rate and substrate temperature can be altered independently [16, 17]. Furthermore OVPD® offers an additional deposition parameter, the deposition pressure, thus in total three deposition parameters to control and modify the morphology of organic films [24].

In first approximation, OLED devices require smooth films and smooth organic interfaces. To achieve a low CoO for the OLED production the tact-time should be short thus the deposition rate should be high. However, it was observed that with increased deposition rates the roughness of the deposited organic films was increased [16, 22, 23]. Thus, it is crucial to ensure that even for high deposition rates smooth organic films can be deposited. For evaluation α-NPD films with different high deposition rates ranging from 8.2 Å/s up to 87 Å/s were deposited on silicon substrates. The rates and thickness of these films were ex-situ verified by ellipsometry. The surface roughness of these 50-nm thick organic films was determined by AFM measurements and the results are shown in Fig. 8.5. The RMS-roughness of these films varies within 0.6 nm and 1.0 nm whereas the RMS value for 87 Å/s is 0.7 nm. These RMS values are in the order of the dimensions of the used organic molecules. This, proves, that even for deposition rates as high as 87 Å/s monomolecular smooth films of α-NPD were deposited, see Fig. 8.5 [25].

Whereas for OLEDs smooth interfaces are preferred the requirements for OPV cells are different. For example, to enhance the OPV efficiency the concept of a bulk heterojunction was introduced to extend the interface responsible for the electron hole splitting. As a consequence a rough morphology composed from well-defined polycrystalline or even crystalline layers are preferred instead of smooth amorphous films as preferred for OLED stacks [20,26,27]. Besides getting a rough morphology on a smooth substrate, also the planarization of a rough substrate can be an issue of interest.

Beside the topology of the organic film also the physical properties can be affected by the optimisation of the deposition parameters. Figure 8.6 shows the measured current density for a hole-only-device as function of the substrate temperature during deposition of the HIL material. Increasing the substrate temperature towards 40°C results in a significant increase of the current density

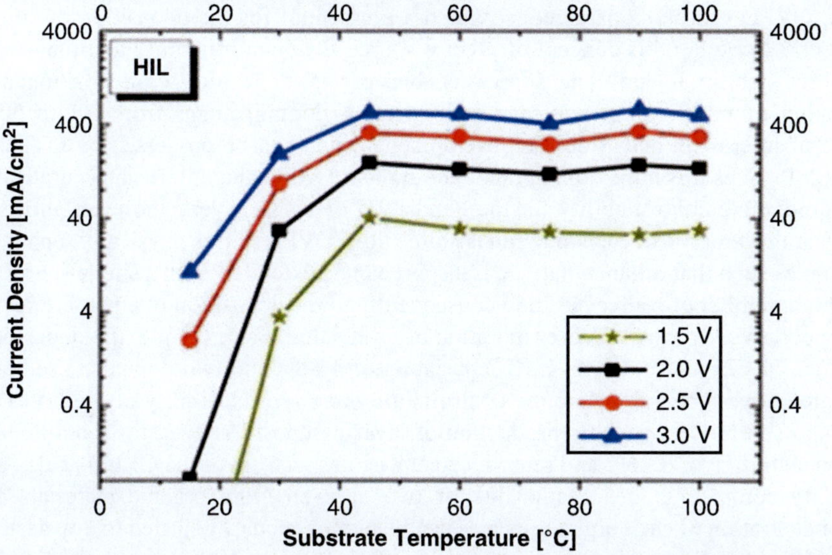

Fig. 8.6 Current density of a hole-only-device based on a material for a hole injection layer (HIL). When deposited at higher substrate temperature the current density improves

improving the electrical conductivity of the investigated HIL material [25]. This demonstrates, that the potential of an organic material, which shows, e.g., at the first investigation no promising material properties, can be fully exploited by optimizing the deposition parameters. This parameter optimisation is obviously not limited to the substrate temperature but also the deposition pressure or rate could be used for optimisation.

Typically the used organic materials offer a broad deposition parameter window with respect to rate, substrate temperature, and pressure for the deposition of smooth films. However, some organic film and material properties might by possible to be improved with respect to the device and production requirements. This is especially true for OPV application, where rough and even crystalline films are preferred for some applications. It can be concluded that OVPD® offers the option to achieve specific organic film morphologies required by individual process parameters.

8.3.3 OLED Stack Designs Fabricated by OVPD® – Cross-Fading

Research on organic semiconductor materials has shown remarkable progress with the introduction of material combinations using hosts and single or multiple guests or additional host materials [9, 13, 28]. In the previous chapter we have explained the OVPD® process parameters, which are essential for the deposition of single layers. Applying this concept of precise source flow controllability and long-term source stability to additional sources enables precise deposition, co-deposition, and doping of multi-component layers. In general, doping ranges from 0.1 to 50% are of interest in actual device development which can be processed by OVPD® [16]. Besides this huge doping range the standard MFC guarantees an accurate and reproducible controllability for the deposition of single layers and even allows a dynamic variation of each individual source flow. OVPD® uses physically separated sources such that organic material transport can be extended from a single source to a high number of sources leading consequently to the deposition of any mixtures or doped layers, known as co-evaporation or co-hosting. For example, the deposition of an OLED stack with an EML consisting of a host and two dopants, a red and a green dye molecule. Here the controllability and reproducibility of the OVPD® process is proven by the reproduction of layer composition even for a red dopant concentration of 0.26% and reproduced emission spectrum of the OLED [20].

By controlling the organic flux of two or even more organic materials the concentration of each organic species can be homogenously adjusted to any desired ratio in an abrupt or a ramping mode offering the potential to create precise constant mixtures or mixtures with a gradual concentration within the layer [16]. Figure 8.7 gives an overview of different potential constant or gradual material mixtures controlled by the deposition rates of, e.g., three materials (r_1, r_2, and r_3). The organic materials are mixed in the gas phase and homogeneously deposited over the substrate due to the CCS technology.

In zone 1, the source flow of materials 1 and 2 is switched on and the total deposition rate is the sum of both individual rates ($r_{tot} = r_1 + r_2$). In zone 2, the

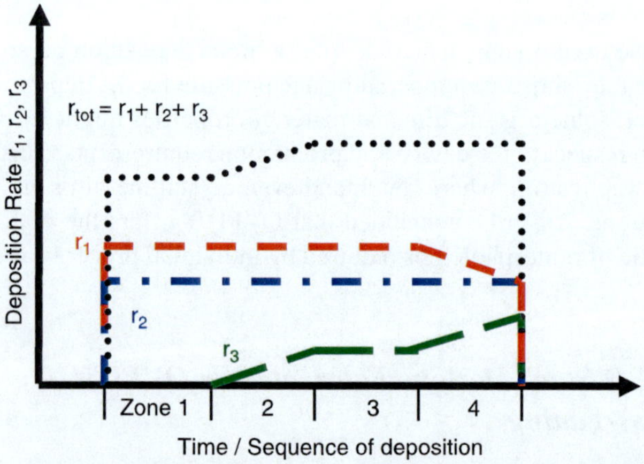

Fig. 8.7 Each zone represents a schematic material mixture based on three organic materials. Zone 1 is a mixture of the materials 1 and 2 and zone 3 of the materials 1 and 2 and 3. In zone 2 the rate r_3 of the material 3 is ramped leading to a concentration gradient. Whereas in zone 4 the rate r_1 is reduced and simultaneously the rate r_3 increased which is described as layer cross-fading [29]

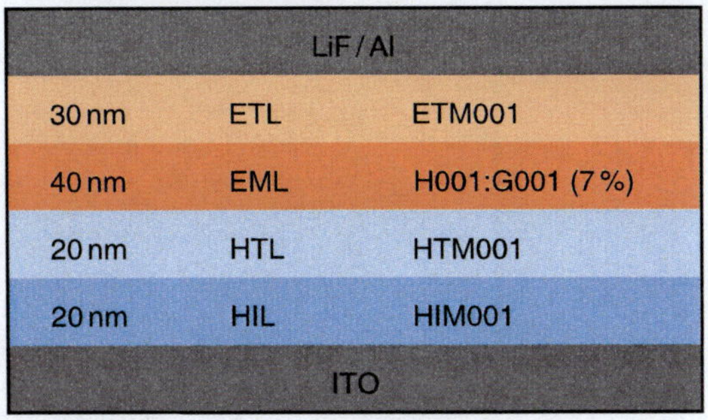

Fig. 8.8 Reference OLED structure with phosphorescent red dopant [29]

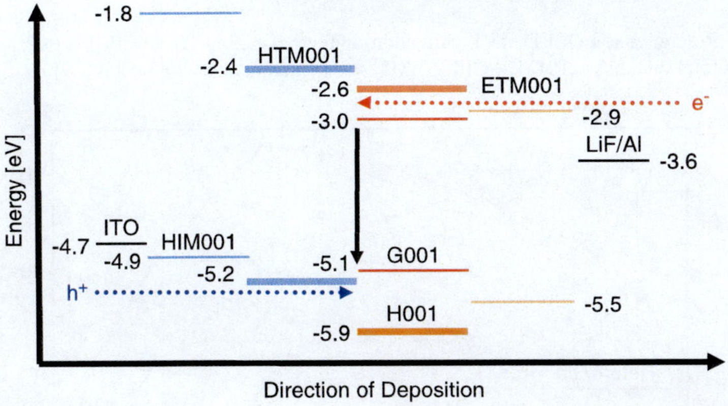

Fig. 8.9 HOMO and LUMO schematic for the reference red OLED, here without the CFL layer. The electrons are injected from the ETL towards the EML and the holes from the HIL via HTL towards the EML. At the interface HTL to EML a charge accumulation occurs [29]

rate of material 3 is increased linearly increasing the total deposition rate ($r_{tot} = r_1 + r_2 + r_3$) and simultaneously the concentration of material 3 in the deposited layer. This leads to a concentration gradient. In zone 3, a constant mixture of the materials 1 and 2 and 3 is deposited. In zone 4, the deposition rate of material 2 is kept constant, whereas material 1 is linearly "faded out" and rate of material 3 is increased with the opposite slope. This special version of a concentration gradient is an example for cross-fading.

The benefit of a concentration gradient on the device performance was already reported either for OLEDs [15, 30] or for OPV fabricated by OVPD® [26]. How the concept of cross-fading can be applied to an organic device and enables a new parameter of freedom for the overall stack and design and will be demonstrated in the following [25, 30–32]. Figure 8.8 shows a reference red OLED stack with 40 nm EML and 20 nm HTL which serves as reference for this investigation. The corresponding HOMO and LUMO levels are given in Fig. 8.9. Due to the gap in the HOMO levels at the HTL to EML interface a charge accumulation is present which is especially relevant at larger currents leading to the well-known roll-off-effect, which is a reduction of the current efficacy as function of the luminance [25, 31].

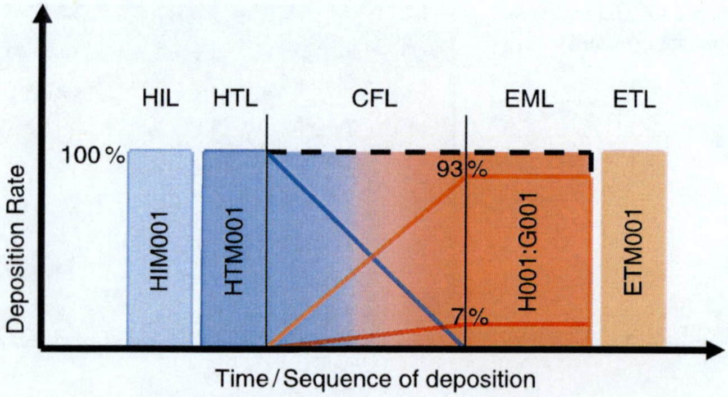

Fig. 8.10 Reference red OLED stack with additional cross-fading layer (CFL) between HTL and EML. The total thickness of HTL+CFL+EML was kept constant with 60 nm [29]

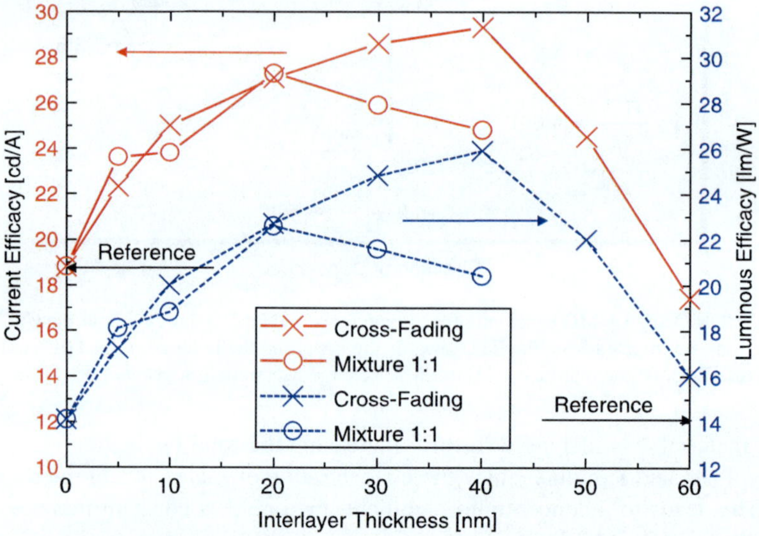

Fig. 8.11 Current and luminous efficacy vs. thickness of CFL and constant mixture interlayers [29]

The cross-fading concept was applied for the red reference OLED by introduction of an additional cross-fading layer (CFL) between HTL and EML as illustrated in Fig. 8.10 [25, 29, 31]. In the CFL is the rate of the hole transport material (HTM001) linearly reduced towards zero whereas the rate of the Host material (H001) and of the red guest respectively red dopant (G001) are linearly increased starting from zero. The total thickness of HTL+CFL+EML was 60 nm and kept constant throughout these experiments. In addition and for comparison a structure with an interlayer between HTL and EML with constant ratio of 1:1 for the HTM001 and H001 were also investigated. The impact of the CFL layer with varying thickness between HTL and EML on the OLED performance is shown in Fig. 8.11 [29].

For this investigation, the reference phosphorescent red OLED device had at 1,000 nits a luminous efficacy of 14.1 lm/W and a current efficacy of 18.8 cd/A at 4.18 V [25, 29, 31]. The color point of all investigated OLEDs either with or without CFL is the same with CIE (0.62/0.38). The motivation for the cross-fading approach is the reduction of charge carrier accumulation at the interface of HTL to EML and a broadening of the recombination zone at higher current densities, thus an improved carrier distribution in the EML. Thus, at the interface HTL to EML an additional layer will be introduced consisting either of a constant mixture of the neighboring organic materials or a cross-fading composition, as illustrated in Fig. 8.10. To ensure an objective judgment and comparison on the achieved the total thickness of HTL+CFL+EML was kept with 60 nm the same as for the reference red OLED stack, see Fig. 8.8. Figure 8.11 shows the measured current efficacy (left axis) and luminous efficacy (right axis) vs. the interlayer thickness in comparison of devices utilizing cross-fading (red) and layers with constant mixing ratio (blue). Increasing the interlayer thickness improves for the layer with a 1:1 mixture as well as for the CFL structure the current efficacy as well as the luminous efficacy. However, at an interlayer thickness of 20 nm the efficacy for the mixture structure reaches its maximum and decreases with increasing interlayer thickness. In contrast the efficacy for the CFL structure improves further. The maximum efficacy is achieved at a CFL thickness of 40 nm with a current efficacy of 29.3 cd/A and luminous efficacy of 25.9 lm/W [25, 29, 31]. Thus, by using the CFL concept the luminous efficacy could be improved by 139% compared to the reference OLED. The increase in efficacy is attributed to an improved hole injection from the HTL to EML suppressing charge accumulation due to the lower HOMO level of the H001 compared to the HTM001. In addition, cross-fading of the predominately hole conducting zone with the predominately electron conducting zone enables a interpenetrating molecular network resulting in a broadening of the recombination zone, especially a higher current densities [29, 31]. As a consequence, the roll-off effect improves due to the CFL approach [25, 31]. Further increase of the CFL thickness beyond 40 nm results in decreasing efficacies attributed to a shift of the recombination zone. It can be concluded that the introduction of an interlayer with a mixture and even with a cross-fading concept are options to enhance the overall device performances [25, 29, 31, 33].

In the above example, an additional CFL was introduced. However, the cross-fading concept can be extended to the EML design itself. Additional optimisation of the red OLED stack lead to 34 lm/W and an EQE of 18% without applying no ILO techniques [29]. Furthermore, the concept of cross-fading can be extended even to an OLED stack with two or more optical dopants. For example, using a phosphorescent green and red dopant in a cross-faded EML enables an efficient and easy adjustable OLED emission ranging from red via yellow towards green [32]. The concept of creating a virtual host material by mixing a predominately hole conducting host with a predominately electron conducting host creating a host with locally preferential electrical properties. This leads to an efficient carrier balancing in the emission zone. Nearly luminance independent CIE coordinates were observed and the external quantum efficacy (EQE) was 16.2%, the luminous efficacy 32.8 lm/W and the current efficacy 36.5 cd/A at 1,000 nits. Applying an EML with a constant mixture of the two host materials does not achieve such results [32]. The limiting factor for the EQE is likely related to the applied phosphorescent green dopant. Further improvements on the yellow OLED stack by improved cross-fading are promising. Thus, the cross-fading concept truly is a measure to improve the OLED performance of monochrome but also mixed OLEDs.

In summary, the realization of vertical doping profiles like simple gradients or CFLs and multi-material mixtures allow a precise layer fine-tuning to optimize device performance. Here, OVPD® combined with the CCS technology offers the ability to fabricate such kind of OLED stacks either in R&D or production and this with high uniformity and further benefits offered by the OVPD® technology.

8.4 Conclusion

The OVPD® technology uses a carrier gas to transport the evaporated organic materials from the source area towards the substrate. This is in contrast to the VTE technology. Beside the pure deposition of the organic film also the morphology and physical properties are crucial, which can be altered and controlled by the OVPD® deposition parameters rate, substrate temperature, and process pressure. A detailed comparison of the OVPD® technology with the VTE technology is given elsewhere [16]. OVPD® combines the individual control of deposition rate of organic materials, their homogenous mixing in the gas phase and deposition as film on the substrate with the option of multi-layer deposition with different layer compositions and even complex concentration gradients like cross-fading. Applying the cross-fading concept to monochrome and multicolor OLED shows significant improvements in the device performance. Thus, in parallel to the replacement of MBE by MOCVD the OVPD® technology has the potential to overcome the limitation of VTE.

Acknowledgments This work was the result of a team effort from the groups at AIXTRON AG, Philips BU OLED, Aachen, Technische Hochschule (TU) Braunschweig, and the Rheinisch Westfälische Technische Hochschule (RWTH), Aachen, which performed most of the work reported in this article.

Part of this work was financially supported by the Federal Ministry of Education and Research in Germany (BMBF, No. 001BD153; No. 13N8650, No. 13N8993, and BMU 0329927B).

OVPD® technology has been exclusively licensed to *AIXTRON* from *Universal Display Corporation*, Ewing, NJ, USA, for equipment manufacture. OVPD® technology is based on an invention by Professor Stephen R. Forrest et al. at Princeton University, USA, which was exclusively licensed to UDC, AIXTRON, and UDC have jointly developed and qualified OVPD® pre-production equipment.

References

1. C.W. Tang, S.A. VanSlyke, Appl. Phys. Lett. **51**, 913 (1987)
2. L.S. Hung, C.H. Chen, Mater. Sci. Eng. **R 39**, 143 (2000)
3. S.R. Forrest, Chem. Rev. **97**, 1793 (1997)
4. S.R. Forrest, MRS Bull. **30**, 28 (2005)
5. B.D'Andrade, V. Adamovich, R. Hewitt, M. Hack, J.J. Brown, SPIE **5937**, 87 (2005)
6. J. Shinar, *Organic Light-Emitting Devices: A Survey*, (Springer, New York, USA, 2004)
7. http://www.lumiblade.com; http://www.osram-os.com
8. J. Kido, M. Kimura, K. Nagai, Science **267**, 1332–1334 (1995)
9. M. Hack, *International Summer School on OLEDs, Organic Electronics: from Lab to Home*, Krutyn, Poland, 2–8 June 2009; Organised by oled100.eu
10. http://www.universaldisplay.com
11. S. Reineke, F. Lindner, G. Schwartz, N. Seidler, K. Walzer, B. Lüssem, K. Leo, Nature **459**, 08003 (2009)
12. H. Kanno, Y. Hamada, N. Matsusue, H. Takahashi, R. Nishikawa, K. Mameno, *SPIE Conference on Organic Light-Emitting Materials and Devices VII*; San Diego, California, USA, 4–6 Aug 2003
13. B.W. D'Andrade, J.Y. Tsai, C. Lin, M.S. Weaver, P.B. Mackenzie, J.J. Brown, *SID 2007* (Long Beach, CA, USA, 20–25 May 2007)
14. N. Meyer, M. Rusu, S. Wiesner, S. Hartmann, D. Keiper, M. Schwambera, M. Gersdorff, M. Kunat, M. Heuken, W. Kowalsky, M.C.h. Lux-Steiner, Eur. Phys. J. Appl. Phys. **46**, 12506 (2009)
15. J. Kido, *Plastic Electronic 2007* (Frankfurt, Germany, **7**.10.n2007)
16. N. Meyer, M. Heuken, *Review About Organic Vapor Phase Deposition for Organic Electronics*, ed. by H. Klauk (Wiley-VCH, Weinheim, Germany, 2006) ISBN: 9783527312641, pp. 203–232
17. B. Marheineke, SPIE Proc. **5961**, 3 (2005)
18. M. Shtein, H.F. Gossenberger, J.B. Benzinger, S.R. Forrest, J. Appl. Phys. **89**(2), 1470 (2001)
19. P.E. Burrows, S.R. Forrest, L.S. Sapochak, J. Schwartz, P. Fenter, T. Buma, V.S. Ban J.L. Forrest, J. Crystal Growth, **156**, 91 (1995)
20. N. Meyer, M. Rusu, S. Wiesner, S. Hartmann, D. Keiper, M. Schwambera, M. Gersdorff, M. Kunat, M. Heuken, W. Kowalsky, M.C.h. Lux-Steiner, Eur. Phys. J. Appl. Phys. **46**(1), 12506, (2009)
21. T. Kato, T. Mori, T. Mizutani, Thin Solid Films **393**, 109 (2001)
22. F. Yang, M. Shtein, S.R. Forrest, J. Appl. Phys. **98**, 014906 (2005)
23. S.Y. Yang, K. Shin, C.E. Park, Adv. Funct. Mater. **15**, 1806–1814 (2005)
24. P. Niyamakom, Phd thesis, RWTH, Aachen, Germany, 2008
25. Project report. Förderkennzeichen BMBF 13N8993 (2009) http://edok01.tib.uni-hannover.de/edoks/e01fb09/6104613891.pdf
26. M. Rusu, J. Gasiorowski, S. Wiesner, D. Keiper, N. Meyer, M. Heuken, K. Fostiropoulos, M.C.h. Lux-Steiner, *EMRS Conference*, Strasbourg, France, 26–30 May 2008
27. M. Rusu, J. Gasiorowski, S. Wiesner, D. Keiper, N. Meyer, M. Heuken, K. Fostiropoulos, M.C.h. Lux-Steiner, (2008): 23rd EUPVSEC, Valencia, Spain, 1–5 September, 2008: WIP-Renewable Energies, ISBN: 3–936338–24–8, pp. 679–681
28. H. Hoppe, N.S. Sariciftci, J. Mater. Res. **19**(7), 1924–1945 (2004)
29. C. Himcinschi, S. Hartmann, A. Janssen, N. Meyer, M. Friedrich, W. Kowalsky, D.R.T. Zahn, M. Heuken, J. Cryst. Growth **275**, e1035 (2005)
30. M. Schwambera, D. Keiper, N. Meyer, M. Heuken, F. Lindla, M. Bösing, C. Zimmermann, F. Jessen, H. Kalisch, R.H. Jansen, P.v. Gemmern, D. Bertram, IMID 2009, KINEX Seoul, Korea, 13–15 October 2009

31. M. Bösing, C. Zimmermann, F. Lindla, F. Jessen, P. van Gemmern, D. Bertram, D. Keiper, N. Meyer, M. Heuken, H. Kalisch, R.H. Jansen, 2009 *Material Research Society (MRS) Spring eeting*, San Francisco, CA, USA, 13–17 April 2009
32. F. Lindla, M. Bösing, C. Zimmermann, F. Jessen, P. van Gemmern, D. Bertram, D. Keiper, N. Meyer, M. Heuken, H. Kalisch, R.H. Jansen, *Material Research Society (MRS) Spring Meeting*, (San Francisco, CA, USA, 13–17 April 2009)
33. F. Lindla, M. Bösing, C. Zimmermann, F. Jessen, P. van Gemmern, D. Bertram, D. Keiper, N. Meyer, M. Heuken, H. Kalisch, R.H. Jansen, Appl. Phys. Lett., **95**, 213305 (2009)

Chapter 9
Computational Studies on Organic Electronic Materials

Leonidas Tsetseris

Abstract Advances in the important technological field of organic electronics require a comprehensive understanding of complex physical mechanisms that control the electronic and transport properties of related materials and devices. In this respect, theoretical and computational studies have been established as an indispensable tool for the explanation of available experimental data, or the prediction of novel structures with enhanced functionalities. Here we outline the key concepts behind some of the most widely-used theoretical approaches, with an emphasis on so-called first-principles quantum-mechanical methods and Monte Carlo (MC) simulations. We also present examples of how these approaches are used to describe the electronic properties, processes of charge carrier hopping, and defect formation in prototype organic semiconductors.

9.1 Introduction

Organic electronics is developing rapidly in one of the most important technological fields with applications that either benefit various aspects of everyday life already, or are expected to become available in the near future. Examples [1–4] include, but are not limited to, organic field-effect transistors (OFET), organic thin film transistors (OTFT), organic photovoltaics (OPV), organic light-emitting diodes (OLED), and e-paper. The main advantages for the use of carbon-based materials, like polymers or small molecules (Fig. 9.1), in technology are the low-cost for their production and their flexibility compared to more traditional materials, especially silicon. These types of properties make organic materials ideal for applications in mobile electronic

L. Tsetseris (✉)
Department of Physics, National Technical University of Athens, GR-15780 Athens, Greece
Department of Physics, Aristotle University of Thessaloniki, GR-54124 Thessaloniki, Greece
Department of Physics and Astronomy, Vanderbilt University, Nashville, TN, USA
e-mail: leont@mail.ntua.gr

S. Logothetidis (ed.), *Nanostructured Materials and Their Applications*,
NanoScience and Technology, DOI 10.1007/978-3-642-22227-6_9,

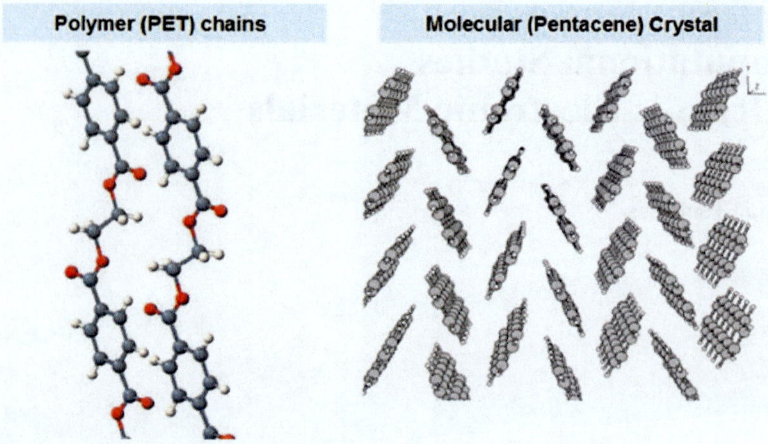

Fig. 9.1 Two examples of prototype materials used in the field of organic electronics. The figure depicts two vicinal poly-ethylene terephthalate (PET) polymeric chains on the left and the herringbone crystalline packing of pentacene on the right

devices, for example cell phones. By the same token, however, traditional electronic materials set standards that their novel organic counterparts need to meet in order to become competitive.

Due to their high technological potential, organic electronic materials and systems have attracted intense and sustained interest in recent years, with scientists employing panoply of experimental approaches for their study. Though several of the requirements for their employment in high-end systems have been met, important challenges remain also open. For example, efforts are continuing to enhance the charge carrier mobilities of organic materials to values comparable to those of amorphous silicon, or the quantum efficiency of carrier generation following light absorption. The resolution of these issues is hampered by the complexity of the underlying physical mechanisms.

Computational studies have emerged as a very effective tool not only for the explanation of available experimental data, but also for the prediction of new materials properties and the design of novel functional systems. The continuous increase of computational power and the development of complex software packages have allowed computational studies on a variety of organic electronic materials and systems that would be very difficult to access just a few years ago.

In this chapter we will describe a number of computational approaches that are commonly employed in today's studies on organic electronic materials. The presentation is by no means supposed to be exhaustive, given the very diverse nature of problems and methods in organic electronics. The aim is rather to offer a brief overview on some of the theoretical techniques often encountered in the literature. For a more extensive presentation the reader can refer to excellent review articles [5–7] on the subject. We will first focus on the so-called ab initio or first-principles approaches that allow the accurate determination of the properties of organic electronic materials, including quantities relevant to carrier transport, with parameter-free calculations. We will then discuss key limitations and extensions of ab initio methods and outline other approaches, such as Monte Carlo (MC) simulations, that enable studies of larger-scale systems. In the second part we will use examples which make clear how the above approaches are being used in today's research to study the properties of organic electronic materials. We will conclude with a brief outlook and perspective for the field of computational organic electronics.

9.2 Computional Methods

9.2.1 A Brief Overview

The problems related to the operation of organic electronic devices span multiple scales. Therefore, in order to perform comprehensive studies in this field there is the need to develop and employ respective computational methods that can deal with the various relevant scales, starting from the atomic-level details of the electronic properties of organic electronic materials, all the way up to device-level modeling.

To probe the physical properties at the atomic-level one needs to solve the Schrödinger equation that governs the quantum-mechanical evolution of electrons and holes. Unfortunately, the exact solution of Schrödinger's equation is impossible for extended systems like molecules or crystals. The Coulomb interactions between the electrons couple their motion and lead to a set of differential equations that cannot be solved even with the most powerful computers. For this reason, various approximations have been adopted leading to computational schemes that are commonly described as ab initio or first-principles methods. The most successful and widely used of these methods is Density-Functional Theory (DFT), which we will describe in more detail below.

Methods like DFT provide very powerful tools in the field of organic electronics. Their range of applicability, however, is limited by the size of the system of interest given that today's computers can typically deal with the quantum-mechanical equations of motion of systems with up to at most a few hundred atoms. Significant limitations exist also with respect to the temporal evolution of systems. Many of the typical time-scales that are relevant to organic electronics are orders of magnitude larger than what can be studied at present with quantum-mechanical approaches. One way to overcome these spatial and temporal limitations is to employ so-called force fields that employ an harmonic or quasi-harmonic description of the interactions between the atoms of the system. The force-field approximation is less accurate than first-principles approaches, but allows molecular dynamics (MD) simulations for systems of hundreds or thousands of atoms at finite temperatures.

Still larger scales, however, are often encountered for important problems like transport of carriers in realistic device geometries. For these types of problems MC simulations are employed to study the charge or energy transfer between different units of a device. In addition to being the suitable tool for larger scales, a MC approach can generically incorporate the disorder that is usually present in the relatively soft materials of organic electronics. Finally, at even larger scales systems can be modeled with rate-equations that describe the flow of currents at the device and circuit level.

Based on the above it is clear that comprehensive computational studies in the field of organic electronics require the synergistic employment of various methods. It is, for example, typical for MD calculations to use force fields that are extracted from higher-level ab initio simulations. MC studies can also take as input parameters derived from first-principles calculations and, in turn, can provide values, such as diffusivities to continuum-based modeling of rate-equations. Depending on the details of implementing this link between different methods one can have the cases of the so-called sequential or concurrent multiscale modeling. In the former case the different methods are employed in sequence, for example one gets first ab initio parameters and then performs MC simulations. In the latter case there is a continuous feedback between different methods that are executed simultaneously. Concurrent multiscale modeling is significantly more challenging to implement, but has the potential of higher accuracy.

9.2.2 First-Principles Methods

The Schrödinger equation is a differential equation of the form

$$H\Psi = E\Psi \tag{9.1}$$

where Ψ and E are the wavefunction and energy of the system. H is the Hamiltonian operator for the system and comprises different terms [8]

$$H = \overbrace{-\sum_i \frac{\hbar^2}{2m_e}\nabla^2_{r_i}}^{A} \overbrace{-\sum_{iI} \frac{Z_I e^2}{|R_I - r_i|}}^{B} \overbrace{-\frac{1}{2}\sum_{ij,i\neq j} \frac{e^2}{|r_j - r_i|}}^{C} \overbrace{-\frac{1}{2}\sum_{IJ,I\neq J} \frac{Z_I Z_J e^2}{|R_I - R_J|}}^{D} \tag{9.2}$$

which describe the kinetic energy (term A) of the electrons, the Coulomb interaction between electrons and ions (B), the Coulomb interaction between the electrons (C), and the interaction between ions (D). This form of the Hamiltonian presupposes the so-called Born-Oppenheimer approximation that neglects the quantum-mechanical equation of motion for the ions and keeps only the one for electrons. This approximation is reasonable because of the large difference between the masses of electrons and nuclei. Electrons are much lighter and, for this reason, they move much faster than the ions so that the latter can be regarded as practically idle at each particular instance.

As we noted above, it is impossible to provide an exact solution of (9.1) for extended systems. The reason is the presence of term C that couples all the electronic degrees of freedom and renders the quantum-mechanical problem intractable. Various approximations to circumvent this problem have been devised. One of the simplest schemes [8] is the so-called Hartree approximation. It describes the electron–electron interactions only at the mean-field level by replacing for each electron the sum of coupled terms C with the term for the interaction between the electron and the field created by all the other electrons in the system. If the Hartree wavefunction is antisymmetrized to account for the fermionic character of the electrons then the approximation is called Hartree-Fock.

The description of the electron–electron interactions at the Hartree-Fock level neglects a very significant part of the Coulomb terms of the type C. This omitted part is commonly referred to as the exchange and correlation interactions of the electrons. Many-body theory techniques, like the so-called Configuration Interaction (CI) or Coupled-Cluster (CC) methods [9], are accurate, but laborious approaches to deal with aspects of the electronic exchange and correlation. A simpler method, that has proven to be very successful in a wide-range of problems of condensed matter physics and quantum chemistry, is DFT.

DFT is based on the Kohn–Hohenberg theorem [10] according to which the electronic charge density of the ground state of a system is uniquely defined by the potential exerted on the electron system by other sources, including the atoms. As a result, there is one-to-one correspondence between the potential and the charge density. Given that the electronic wavefunction depends on the potential term in the Schrödinger equation, the one-to-one correspondence indicates that the wavefunction is also determined uniquely by the density. And since the total energy of the system depends on the wavefunction, we can conclude that the energy of the ground state can also be written as a functional of the density

$$E = E[n(r)] \tag{9.3}$$

The key challenge in DFT is to determine the energy functional and use it to obtain the wavefunctions and the properties of the system. By varying the functional one can find the energy minimum and obtain thus the properties of the ground state of the system. A common approximation employed at this point is the one that leads to the Kohn–Sham set of equations [11]

$$\left[-\nabla^2 + V^{\text{eff}}\right]\varphi_i(r) = \varepsilon_i \varphi_i(r) \tag{9.4}$$

for single-particle orbitals ϕ_i. In (9.4) the effective potential V^{eff} is given by the expression [8]

$$V^{\text{eff}}(r) = V(r) + e^2 \int \frac{n(r')}{|r - r'|} dr' + \frac{\delta E^{xc}[n(r)]}{\delta n(r)} \tag{9.5}$$

where $V(r)$ is the external potential, the second term gives the mean Coulomb field due to all the other electrons, and the last term (which we will call V_{xc}) represents the exchange-correlation effects. The central question in the development of DFT is the construction of the last exchange-correlation potential term. One common approximation, known as the Local Density Approximation (LDA), is to assume that V_{xc} depends only on the value of the density at a particular site, i.e., $V_{xc}(r) = V_{xc}[n(r)]$. One can try to improve LDA functionals by allowing V_{xc} to depend not only on the value of $n(r)$ at a particular site, but also on its gradient. In this way one can construct what are known as Generalized-Gradient Approximation (GGA) exchange-correlation functionals.

There are a number [4, 5, 8] of well-known LDA and GGA functionals that are used widely in solid-state physics and quantum chemistry. They can be constructed by fitting DFT-based results to those of many-body theory approaches or, in more recent years, to accurate numerical methods such as Quantum MC techniques. Though there are important limitations for the validity of the functionals, as we will discuss in more detail below, DFT generally provides an accurate prediction of several materials properties. For example, it is typical that DFT predictions for lattice constants of inorganic crystals fall within 1–2% of the experimental values. We will present below examples of successful application of DFT in studies on organic electronic materials and cases that need additional consideration to bring the calculated results closer to measured data.

9.2.3 First-Principles Methods: Limitations and Extensions

One of the best-known limitations of DFT studies is the underestimation of the energy separation between occupied and unoccupied electronic states. In the case of finite systems this separation is called the HOMO-LUMO gap, where HOMO (LUMO) refers to the Highest Occupied (Lowest Unoccupied) Molecular Orbital. For extended systems the separation corresponds to the energy band gap. Typical differences between calculated and measured energy gaps are in the range of several tenths of eV or even larger than an electron-volt. For example, the calculated energy band gap for pentacene crystals, as we will show below, is about 0.8 eV, whereas the experimental value is more than 2 eV.

The exact reason, or reasons, of the failure of DFT functionals with respect to the HOMO-LUMO gap is currently under debate. The treatment of screening is known to be important in this respect and methods that include post-DFT corrections for screening find larger band gaps that are closer to the experimental data. One such approach is the GW method [12]. The name of the approach represents the self-energy correction for quasiparticles expressed through the Green's functions G and the screened interaction W.

Another important limitation of popular DFT functionals is their neglect of van der Waals interactions. This omission is not critical for solid-state inorganic materials where the nature of bonding (covalent, metallic, or ionic) does not depend strongly on the long-range dispersive forces. In contrast, for organic materials the van der Waals interactions is a non-negligible component of the cohesive forces. The ideal way to correct for the neglect of van der Waals forces is to include them in the DFT functional. Efforts in this direction have been under way [13] in recent years. A simpler approach is to include van der Waals interactions as post-DFT corrections. The usual empirical way to describe dispersive forces is through potentials of the Lennard-Jones type

$$E_{\mathrm{dis}} = \frac{A}{R^{12}} - \frac{C}{R^6} \tag{9.6}$$

where the first term describes the exchange-related repulsion and the second term the long-range inter-atomic attraction. Since DFT functionals take into account exchange, but not the long-range dispersive forces, a van der Waals correction should be added without destroying the short-range description of DFT. This can be achieved by using an envelope function to write the dispersion correction in the form [14]

$$E_{\mathrm{dis}} = -f(R)\frac{C}{R^6} \tag{9.7}$$

A suitable such envelope function is depicted in Fig. 9.2.

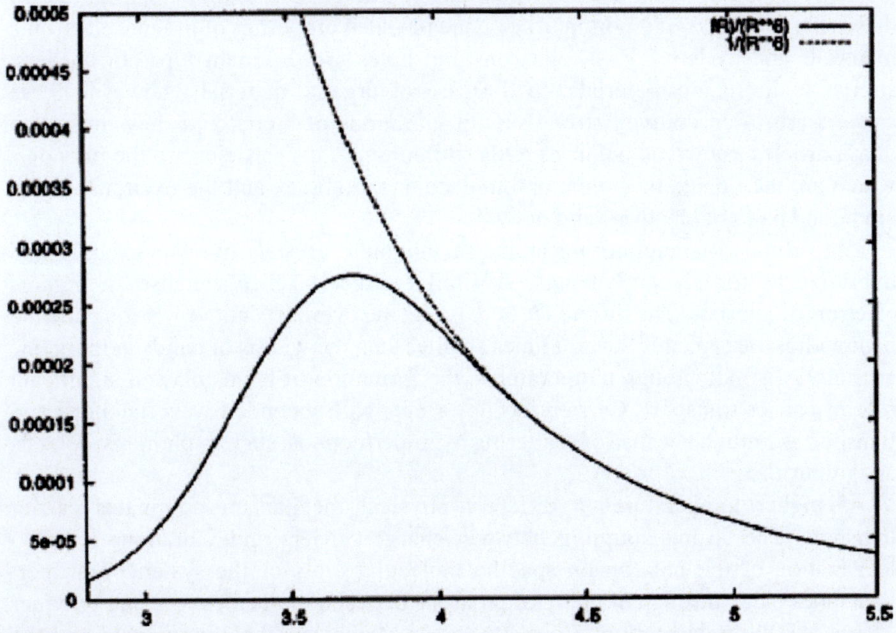

Fig. 9.2 The attractive part of a Lennard-Jones potential and the same function modified at small distances by an appropriate envelope function

In addition to generic limitations of today's DFT functionals, there are often issues with the computational power needed to perform calculations relevant to the operation of organic electronic devices. We mentioned above that the simulation box of ab initio studies is typically limited to a few hundreds of atoms, size that is too small to study long-range transport and other interesting phenomena. DFT calculations on electron–phonon coupling or exciton dynamics, processes that determine transport and optoelectronic properties, are cumbersome and can be carried out only in the most powerful computing systems.

There are, of course, several other extensions [8] of DFT approaches, some of which are being under development at present. We would like just to mention at this point that time-dependent DFT for the electronic evolution of excited states, density-functional perturbation theory for the phonon spectrum, methods for the determination of minimum energy pathways of transformations and chemical reactions, and the capabilities to calculate local vibrational modes or the dielectric function are important tools that enhance the value of ab initio studies in the field of organic electronics.

9.2.4 Carrier Hopping Mechanisms

Because the quantum-mechanical evolution of molecules and crystals is very complex, the problem is often broken down to specific degrees of freedom that are represented by what are known as quasiparticles or excitations. A number of these species plays an important role on the physical properties of organic electronic materials and systems. First, electrons and holes are the main types of carriers, similar to inorganic materials. In the case of organic materials, however, these carriers are often coupled strongly with vibrations of atoms and they thus form quasiparticles known as polarons. Other important quasiparticles are the phonons, which are the quanta of atomic and molecular vibrations, and the excitons, which are bound pairs of electrons and holes.

The orbitals of neighboring atoms in inorganic crystals overlap strongly and hybridize to form energy bands. A similar effect takes place also in organic electronic materials. The difference here is that the overlap is not very strong and the bandwidths are typically several times smaller than the widths of bands in inorganic crystals. At small enough temperatures, the formation of bands plays a significant role in carrier transport. Carriers occupy states with extended wavefunctions and transport is limited because of scattering by imperfections, such as phonons, defects, and impurities.

At higher temperatures a different physical mechanism dominates carrier transport. The strong coupling between charge carriers and vibrations leads to localization of the polaron on specific molecular units of the system. Transport then takes place through hopping of polarons between molecules or along polymer chains. As the polaron hops from site to site it changes the charge state and the energy of molecular species. This energy variation determines the rate of transfer of the polaron, including its dependence on temperature.

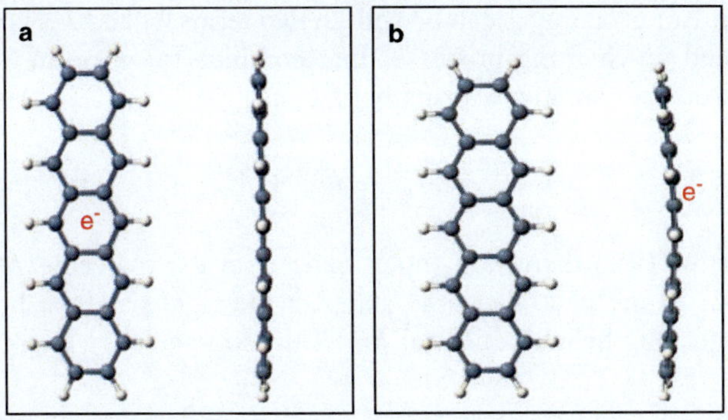

Fig. 9.3 Schematic representation of carrier hopping events between a pair of pentacene molecules. Transfer of the electron from the left molecule in (**a**) to the right molecule in (**b**) leads to coupling with molecular vibrations depicted here as slight bending of the right molecule

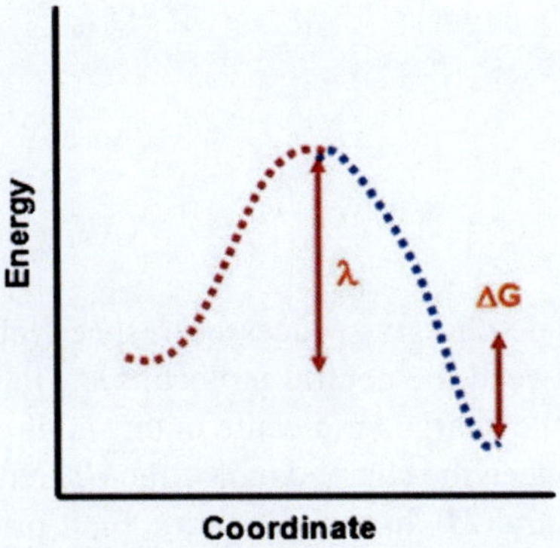

Fig. 9.4 Schematic for the energy variation during a polaron hopping event. ΔG is the free energy difference between the initial and final configurations and λ is the so-called reorganization energy

Figure 9.3 depicts schematically such a transfer process for a pair of pentacene molecules. In Fig. 9.3a the electronic carrier resides on the left molecule of the pentacene dimer, while in Fig. 9.3b it jumps to the right moiety. The transfer of the electron to the right molecule induces deformation related to polaron creation. This polaronic deformation is depicted in Fig. 9.3b as a slight bending of the molecule.

As the schematic of Fig. 9.4 for the energy variation during polaron hopping shows the energy of the system first increases and then relaxes to a new local equilibrium configuration whose energy may differ from that of the original structure by ΔG. The intermediate increase λ is called the reorganization energy, [4–6] a key quantity for the description of polaron transport. An important challenge for computational studies on organic electronic materials is to calculate these values (ΔG and λ) and use them to determine the rate of carrier hopping.

The reorganization energy λ can be split in two terms λ_1 and λ_2 which represent the charging and de-charging processes that combine to make the total hopping event. These processes are given formally [4, 5] as

$$M_a^+ + M_b^0 \rightarrow M_a^0 + M_b^+ \tag{9.8}$$

In the left part of (9.8) there is a (hole) carrier on the molecule M_a, while the molecule M_b is in the neutral charge state. In the right part, the hole moves to molecule M_b, leaving behind a neutral M_a. The reorganization energy λ can then be written as [4, 5]

$$\lambda_{\text{org}} = \lambda_1 + \lambda_2 \tag{9.9}$$

where

$$\lambda_1 = E^{(+)}(M_a) - E^{(0)}(M_a) \tag{9.10}$$

and

$$\lambda_2 = E^{(0)}(M_b^+) - E^{(+)}(M_b^+) \tag{9.11}$$

relate to the charging and de-charging processes, respectively. In particular, λ_1 gives the energy difference between the neutral molecule M_a in its equilibrium configuration and the energy of the charged molecule in this configuration. Likewise, λ_2 is the energy difference between the charged molecule M_b in its ground state geometry and the energy of the neutral M_b in this geometry. First-principles calculations can obtain the total energies of molecules for different charge states and relax geometries to find the equilibrium configurations. The calculated energies can then be used in the above expressions to calculate the reorganization energies λ.

The energy changes described by the reorganization quantities λ are one type of parameters needed to address the hopping rate of charge carriers between molecules or chemical units. Other important quantities are the typical tight-binding parameters that describe orbital overlap and energy band formation in extended systems. Such are the on-site energies [4, 5]

$$\varepsilon_m = \langle \phi_m(r - R_m) \, | H_e | \, \phi_m(r - R_m) \rangle \tag{9.12}$$

and the transfer integrals

$$t_{nm} = \langle \varphi_m(r - R_m) \, | H_e | \, \varphi_n(r - R_n) \rangle \tag{9.13}$$

In the above, ϕ_m are the molecular orbitals centered at sites R_m and H_e is the Hamiltonian quantum-mechanical operator of the system. After the orbitals are determined by solving the Schrödinger equation, the on-site energies and transfer integrals can be obtained by performing the integrations described in the expressions (9.12) and (9.13). Alternatively, the transfer integrals can be obtained by the so-called energy-splitting-dimer [4, 5] approximation

$$t = \frac{\varepsilon_{L+1[H]} - \varepsilon_{L[H-1]}}{2} \tag{9.14}$$

Equation (9.14) provides the t parameters in terms of the on-site energies of the LUMO and the molecular orbital above the LUMO (or the HOMO and the molecular orbital just below the HOMO).

The key question, of course, in the field of organic electronics is to determine the transport properties of related systems. In terms of physical mechanisms, this requirement is equivalent to determining the rate of hopping of the dominant carriers among the chemical units of the system. Hopping can, in fact, be viewed as an electronic transition that takes the electron (or the hole) from a localized state in one unit and transfers it to another localized state on another unit. The way to determine the rate k_{if} of an electronic transition within quantum mechanics is to use the so-called Fermi's Golden rule

$$k_{if} = \frac{2\pi}{\hbar}\left|\langle\psi_i\,|V|\,\psi_f\rangle\right|^2 \rho\left(E_f\right) = \frac{2\pi}{\hbar}\left|V_{if}\right|^2 \rho\left(E_f\right) \tag{9.15}$$

where ψ_i (ψ_f) is the wavefunction of the initial (final) state, V is the interaction that causes the transition, and $\rho(E_f)$ is the density of states (DOS) at the energy E_f of interest.

Starting from (9.15) and using suitable approximations one can find expressions that give the rate of carrier hopping in terms of the reorganization energy λ and the change of free energy ΔG during a hopping event. In the case of high temperature and strong coupling between the electronic and vibrational degrees of freedom this relation is known as the Marcus expression [4–6]

$$k_{if} = \frac{t^2}{\hbar}\sqrt{\frac{\pi}{k_B T\lambda}}\exp\left[-\left(\Delta G + \lambda\right)^2/4k_B T\lambda\right] \tag{9.16}$$

In the other regime of low temperature and weak coupling between electrons and molecular vibrations the hopping rate is given by the so-called Miller-Abrahams expression [4, 5]

$$k_{if} = \nu\exp\left(-2\gamma R_{ij}\right)\begin{cases}\exp\left(-\dfrac{\varepsilon_j - \varepsilon_i}{k_B T}\right), & \varepsilon_j > \varepsilon_i\\ 1, \quad \varepsilon_j < \varepsilon_i\end{cases} \tag{9.17}$$

where ε_i and ε_j are on-site energies, ν is an attempt frequency, R_{ij} is the distance between the i, j sites, and γ is the overlap factor.

9.2.5 Monte Carlo Simulations

Once the rates of hopping between different sites are calculated from one of the above expressions, then the long-range migration of carriers can be simulated with MC techniques. The central idea in this approach is the acceptance or rejection of an individual hopping step based on the Metropolis algorithm. For this purpose one first selects a certain time interval and then calculates the probability that hopping takes place based on the time step and the hopping rate. This calculated probability is compared with a random number with two possible scenarios: if the probability is larger than the random number then the hopping step is accepted and the carrier is placed in the final state of the transition. Otherwise, the hopping attempt is rejected and another hopping step is tested with the algorithm. Alternatively, using the transfer rates and random numbers one can calculate the time required for several different possible hopping steps. The step with the smallest time is then executed and the process is repeated.

The MC technique allows simulations for realistic geometries of organic electronic systems. These geometries may include important elements of electronic devices such as interfaces, but also the generic disorder that is often present in soft materials. The Miller-Abrahams expression, in particular, provides a facile way to incorporate disorder through the variation of on-site energies based on a Gaussian distribution of the form [4, 5]

$$Q(\varepsilon) = \frac{1}{\sqrt{2\pi\sigma^2}} \exp\left(-\frac{\varepsilon^2}{2\sigma^2}\right) \tag{9.18}$$

In the above, $Q(\varepsilon)$ is the probability for a certain on-site energy ε and σ is the standard deviation of the distribution.

We discussed above ways to calculate the rate k_{if} for transition of a charge carrier between orbitals localized on neighboring chemical units. Since this transition is equivalent to a hopping step, the rate k_{if} is related to the constant D of diffusion of carriers $k_{if} = a^2 D$, where a is the hopping distance. The carrier mobility is then given by the expression

$$\mu = \frac{eD}{k_B T} \tag{9.19}$$

where e is the electron charge, k_B is the Boltzmann constant, and T is the temperature. In addition to the hopping-related mobility term, there is a second mobility contribution associated with band hopping. Formally we have

$$\mu = \mu_{\text{tun}} + \mu_{\text{hop}} \tag{9.20}$$

where the first term is the tunneling or band hopping mobility and the second term is the polaron hopping contribution. The physical mechanism that determines the tunneling term is scattering by various imperfections, including phonons. Within the so-called relaxation time approximation, μ_{tun} is given by

$$\mu_{\text{tun}} = \frac{e\tau}{m_{\text{eff}}}. \tag{9.21}$$

μ_{eff} is the effective mass of the carriers and the relaxation time τ represents the time spent between scattering events. In general, the coherent contribution of the tunneling term dominates at small temperatures and for large bandwidths, while the hopping term becomes the controlling factor at higher temperatures where vibrations and their coupling to the electronic degrees of freedom are enhanced. Figure 9.5 shows a schematic [5] for the variation of charge carrier mobility with temperature. The mobility initially decreases rapidly as the temperature increases because of scattering by phonons, but then the polaron part takes over and the mobility grows larger again.

Charge carrier mobilities are key parameters that can be used as input in equations to determine the current-voltage characteristics of organic electronic devices. For example, the switching operation of organic field-effect transistors is described by the familiar FET equations. For a channel of length L and width W

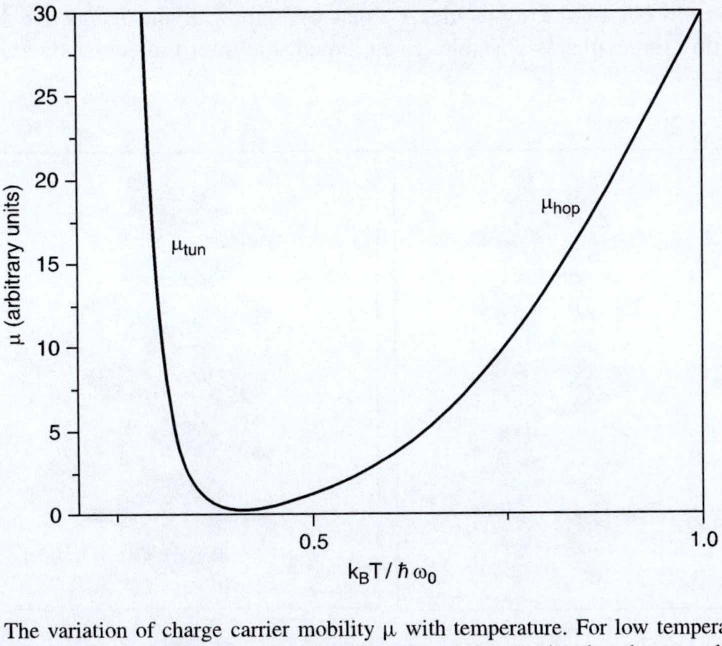

Fig. 9.5 The variation of charge carrier mobility μ with temperature. For low temperatures the tunneling part of the mobility, related to band hopping and scattering by phonons, dominates. At higher temperatures polaron hopping becomes the mechanism that controls the mobility [Adopted from [5]]

and for an applied bias V_{SD} the current I_{SD} between source and drain is given by

$$I_{\mathrm{SD}} = \frac{W}{L}\mu C\,(V_G - V_T)\,V_{\mathrm{SD}}, \tag{9.22}$$

where C is the capacitance of the gate dielectric. In the saturated regime we have

$$I_{\mathrm{SD}} = \frac{W}{L}\mu C\,(V_G - V_T)^2\,. \tag{9.23}$$

9.3 Results and Findings

In this section we will present selective results on the application of some of the methods we discussed above on physical traits of organic electronic materials and devices. The results show how ab initio calculations can be utilized to describe the electronic properties of materials. We also discuss findings based on first-principles methods on the effect of defects and impurities in organic semiconductors. Results on reorganization energies and their use in MC simulations are presented as well.

Figure 9.6 shows the square of the amplitude of the highest occupied and lowest unoccupied molecular orbitals of pentacene. The wavefunctions have lobes that protrude out of the molecules so that overlap with the respective lobes of neighboring molecules is possible. Even though the inter-molecular overlap is not

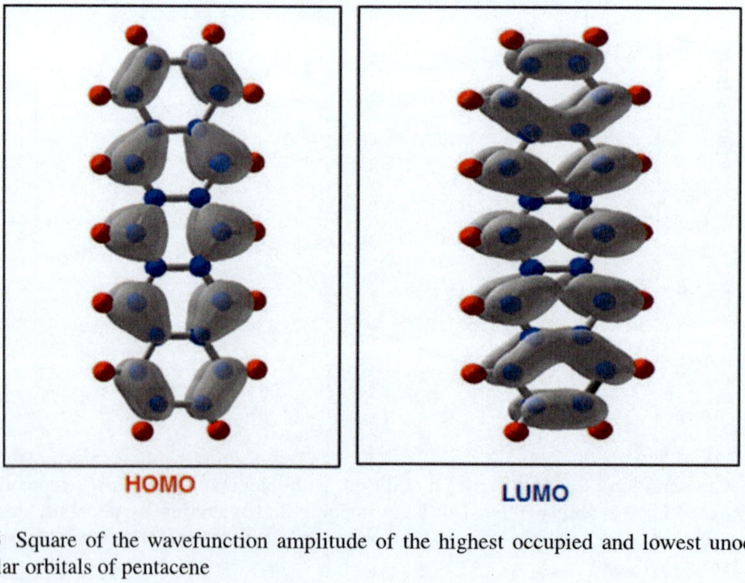

Fig. 9.6 Square of the wavefunction amplitude of the highest occupied and lowest unoccupied molecular orbitals of pentacene

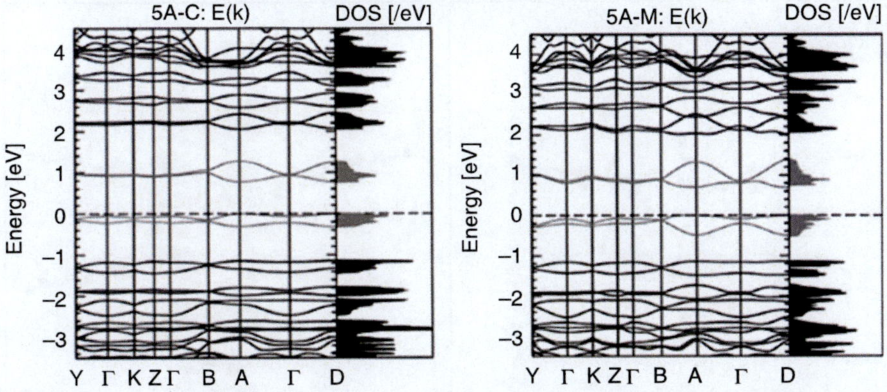

Fig. 9.7 DFT results on the energy band structure and electronic density of states (DOS) for two pentacene crystal polymorphs. The letters in the x-axes represent high symmetry points of the Brillouin zone [Adapted from [15]]

nearly as strong as the overlap within the molecules, it leads to hybridization and, eventually, to formation of energy bands in an extended system.

In Fig. 9.7 we show results [15] on the energy band structure of two pentacene crystalline polymorphs. There are a number of known pentacene polymorphs from experiments. They all exemplify the herringbone structure of Fig. 9.1b, but they differ slightly in the inter-molecular packing and tilting angles. Though these differences are small they have discernible effects in the corresponding band structures. For example, the band width [15] of the valence band is larger in the case of the second polymorph on the right. Differences can also be seen in the higher conduction bands and these differences may affect the optical response of the system.

One thing to note in the band structures of Fig. 9.7 is the value of the calculated energy band gap, about 0.8 eV. This value is much smaller than the experimental band gap of pentacene which is measured [16] at about 2.2 eV. The discrepancy is a typical example of the underestimation of band gap values by DFT studies based on the most popular exchange-correlation functionals. GW calculations succeed in correcting to a large extent this discrepancy. In particular, a GW study [17] on the same pentacene polymorphs as those of Fig. 9.4 gave energy band gap values of 1.8–2 eV.

Figure 9.8 depicts the square of the HOMO and LUMO wavefunction orbitals of rubrene, another prototype organic electronic material that currently holds the record among organic semiconductors in terms of charge carrier mobilities. A rubrene molecule comprises a tetracene backbone and four side phenyl groups. As shown in the figure, the HOMO and LUMO lobes are located mainly on the tetracene backbone, while the contributions from the phenyl groups are very small. Concomitantly, the side groups play only a minor role in the formation of the valence and conduction bands. This diminished role is demonstrated clearly in Fig. 9.9,

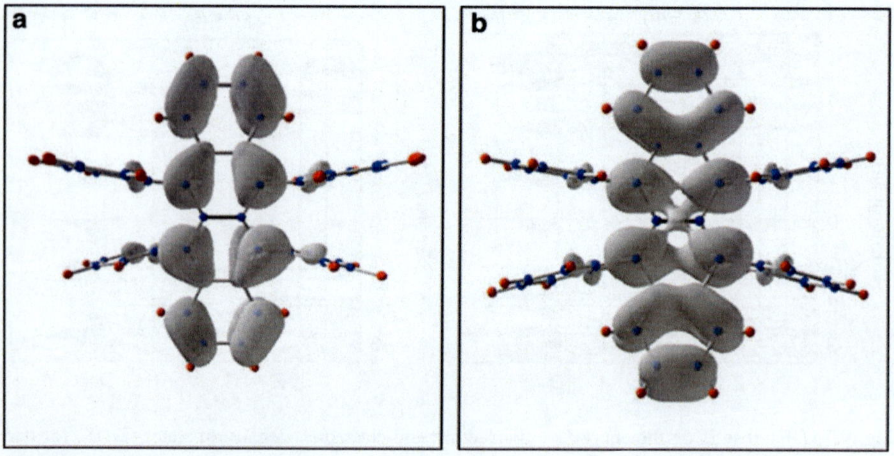

Fig. 9.8 Square of the wavefunction amplitude of the highest occupied and lowest unoccupied molecular orbitals of rubrene

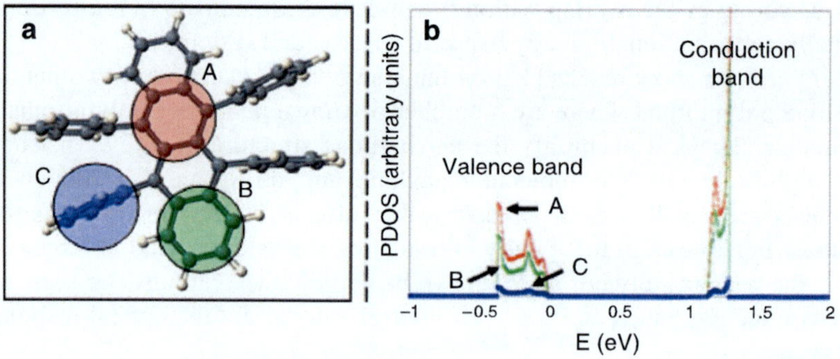

Fig. 9.9 (**a**) A rubrene molecule. (**b**) Projected density of states (PDOS) of crystalline rubrene for the three benzene rings highlighted in (**a**) [Adapted from [18]]

which presents the projected electronic density of states [18] of (PDOS) crystalline rubrene. The PDOS related to the phenyl groups is much smaller than the PDOS associated with benzene rings of the tetracene backbone.

Given the prevalent role of the tetracene backbone in the formation of the valence and conduction band we can infer that polaron hopping happens mainly through these particular chemical units. Moreover, any disruption of the structural and chemical properties of these units is expected to have a large effect on the electronic and transport properties of crystalline rubrene. Indeed, such large effects are obtained [19] in the case of oxygen and water-related impurities in rubrene.

One of the most important issues in the operation of organic electronic devices is their reliability. Organic-based systems typically degrade much faster than their inorganic counterparts, an effect that is accelerated when they are exposed to air or elevated temperatures. Oxygen and water-related impurities are common defect culprits that are believed to give rise to most of the degradation processes in organic devices. The same feature behind the flexibility of organic-based systems, namely their non-covalent inter-molecular bonding, is also responsible for the facile incorporation of impurities. Extrinsic species may enter organic materials either in the initial synthesis and growth stages, or during the long-term operation of related devices. The presence of impurities is particularly detrimental to the operation of organic field-effect transistors when they introduce energy levels in the band gap of the host system. Such levels can act as trapping sites for charge carriers, degrading the mobility and overall transport characteristics.

Ab initio studies [20, 21] on crystalline pentacene have identified a number of hydrogen and oxygen-related configurations which can indeed act as carrier traps. Specifically, a hydrogen adatom adsorbed on a pentacene molecule creates [20] a level that is located about 0.34 eV above the valence band maximum. Interstitial oxygen species, on the other hand, can take several different metastable configurations [21] inside a pentacene crystal, including epoxy structures and inter-molecular bridges of the type shown in Fig. 9.10. The latter oxygen impurity configuration has two levels in the pentacene band gap, as shown in the electronic DOSs plot of Fig. 9.10. The position of these gap states are in close agreement with pertinent observations [22] of oxygen-related carrier traps in pentacene. Oxygen and water-related carrier trap configurations have also been identified with first-principles calculations [19] in the case of rubrene, again in agreement with corresponding experiments [23].

As stressed above, computational studies can play an important role in probing the transport properties of organic electronic materials and, in particular, their dependence on structural and chemical characteristics. One good example is the

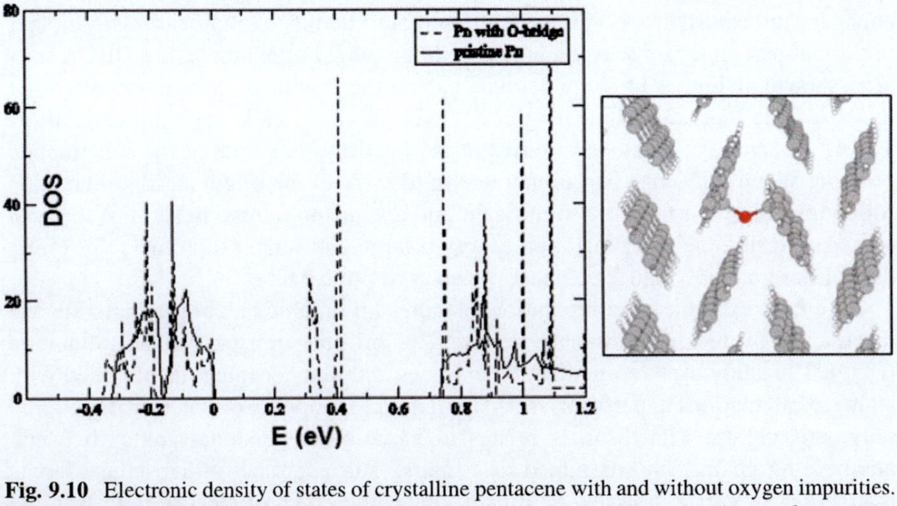

Fig. 9.10 Electronic density of states of crystalline pentacene with and without oxygen impurities. The inter-molecular O bridge on the right creates levels in the energy band gap of pentacene [Adapted from [21]]

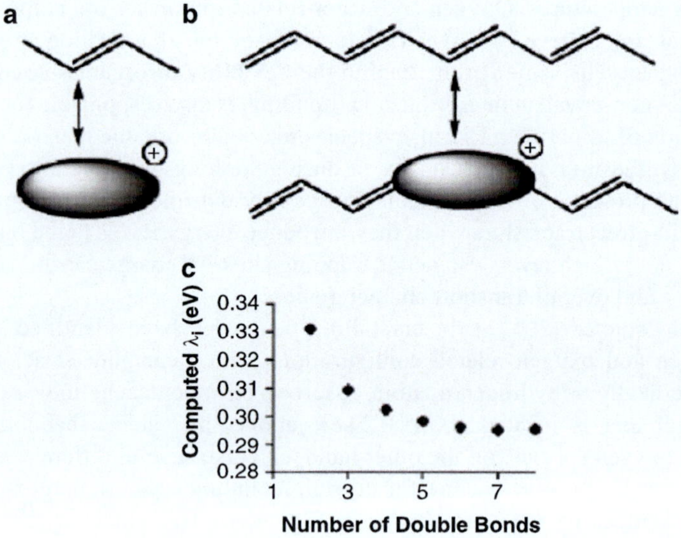

Fig. 9.11 Dependence of calculated reorganization energies on the length of oligomers of polyenes [Adapted from [25]]

determination through ab initio calculations [24] of the reorganization energies of triphenylamine and the TPD diamine, prototype hole-transporting materials that are used in multi-layer OLEDs. The calculated TPD λ value (0.29 eV) is about three times larger than that of triphenylamine, indicating the important role of the central TPD biphenyl group in hindering polaron hopping. Knowledge of the moiety-specific contribution to λ can be very helpful in designing new materials with enhanced transport properties through suitable chemical modifications. Another important parameter for λ is the length of oligomers. For example, the DFT results [25] shown in Fig. 9.11 demonstrate that as the length of a polyene oligomer increases, the reorganization energies decrease monotonically and tend to saturate. The effect can be understood in terms of the localization length of the self-trapped polaron. When this length is comparable to the size of the oligomer, the relaxation following polaron trapping is significant and extends to a large portion of the host unit. As a result, the reorganization energy is large. For longer oligomers this effect diminishes gradually and λ decreases to an asymptotic value.

As a final example of computational studies on organic electronic materials we discuss briefly recent multi-scale results [26] on hole transport in poly-fluorene (PFO). The study uses reorganization energies and transfer integrals obtained with ab initio calculations to perform a MC simulation for hopping of holes on disordered polymeric chains. The disorder relates to variation of torsional angles between neighboring chains, approximated as trimers. The approach utilizes the Marcus expression for carrier hopping and incorporates the effect of electric field **F** in the form of $\Delta G = e\Delta \mathbf{r}_{ij}.\mathbf{F}$, where ΔG is the free energy change following hopping

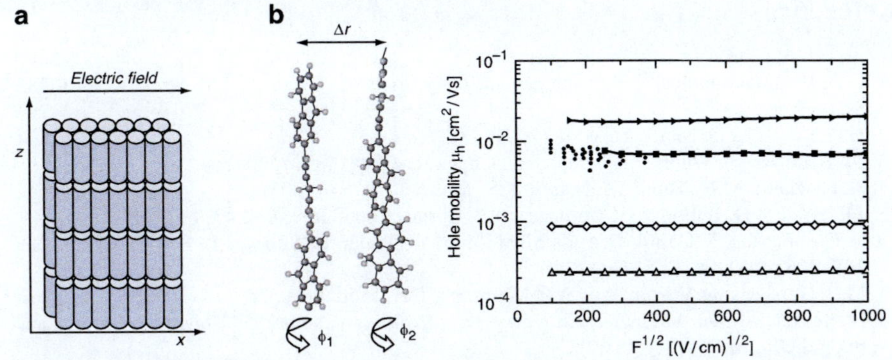

Fig. 9.12 Multi-scale study on hole mobilities of disordered poly-fluorene (PFO). (**a**) depicts the simulation box of PFO trimers employed in the calculations. (**b**) shows the torsional angles that control the rates of hole transfer between neighboring PFO chains. On the right plot a comparison is given between calculated mobilities (lines) and measured data (squares) [Adapted from [26]]

along $\Delta\mathbf{r}_{ij}$. In this way, hole mobilities can be calculated as a function of applied electric field. The results are shown in Fig. 9.12 and they reproduce with satisfactory agreement available experimental data for a certain level of disorder.

9.4 Summary and Outlook

Computational studies on organic electronic materials play an important role in the rapid advancement of this thriving field of technology and basic research. We have described a number of theoretical approaches that are commonly employed in organic electronics to address some of the key questions related to operation of organic-based devices. Ab initio quantum-mechanical calculations provide an accurate and robust description of the physical properties of materials at the atomic-level. These calculations can obtain parameters like reorganization energies and transfer integrals which can then be used in other approaches, for example MC simulations, to address the large-scale characteristics of realistic organic electronic systems.

Simulations have achieved significant successes over the years in explaining key features of polymers and small molecule organic materials. The problems in the field of organic electronics, however, are often quite complex due to, for example, disorder and enhanced electronic correlations. In this respect, there is significant room for improvement in related computational methods. Continuous advancements in theoretical approaches that enhance the accuracy of calculated transport parameters, along with the rapid increase of available computational power and the optimization of related algorithms and codes, strongly suggest that computational studies on organic electronic materials will grow further in the following years as one of the most flourishing fields in research and technology.

References

1. C.D. Dimitrakopoulos, S. Purushothaman, J. Kymissis, A. Callegari, J.M. Shaw, Science **283**, 822 (1999)
2. H.E. Katz, Chem. Mater. **16**, 4748 (2004)
3. M. Bendikov, F. Wudl, D.F. Perepichka, Chem. Rev. **104**, 4891 (2004)
4. A. Facchetti, M.H. Yoon, T.J. Marks, Adv. Mater. **17**, 1705 (2005)
5. J.L. Bredas, D. Beljonne, V. Coropceanu, J. Cornil, Chem. Rev. **104**, 4971 (2004)
6. V. Coropceanu, J. Cornil, D.A. da Silva Fihlo, Y. Olivier, R. Silbey, J.L. Bredas, Chem. Rev. **107**, 926 (2007)
7. M.E. Gershenson, V. Podzorov, A.F. Morpurgo, Rev. Mod. Phys. **78**, 973 (2006)
8. E. Kaxiras, *Atomic and Electronic Structure of Solids* (Cambridge University Press, Cambridge, UK, 2003)
9. J. Paldus, X.Z. Li, Adv. Chem. Phys. **110**, 1 (1999)
10. P. Hohenberg, W. Kohn, Phys. Rev. **136**, B864 (1964)
11. W. Kohn, L. Sham, Phys. Rev. **140**, A1133 (1965)
12. L. Hedin, Phys. Rev. **139**, A796 (1965)
13. M. Dion, H. Rydberg, E. Schröder, D.C. Langreth, B.L. Lundqvist, Phys. Rev. Lett. **92**, 246401 (2004)
14. M. Elstner, P. Hobza, T. Frauenheim, S. Suhai, E. Kaxiras, J. Chem. Phys. **114**, 5149 (2001)
15. K. Hummer, C. Ambrosch-Drexl, Phys. Rev. B **72**, 205205 (2005)
16. E.A. Silinsh, V.A. Kolesnikov, L.J. Muzikante, D.R. Balode, Phys. Stat. Sol. B **113**, 379 (1982)
17. M.L. Tiago, J.E. Northrup, S.G. Louie, Phys. Rev. B **67**, 115212 (2003)
18. L. Tsetseris, S.T. Pantelides, Europ. Phys. J. Appl. Phys. **46**, 12511 (2009)
19. L. Tsetseris, S.T. Pantelides, Org. Electr. **10**, 333 (2009)
20. J.E. Northrup, M.L. Chabinyc, Phys. Rev. B **68**, 041202 (2003)
21. L. Tsetseris, S.T. Pantelides, Phys. Rev. B **75**, 153202 (2007)
22. D.V. Lang, X. Chi, T. Siegrist, A.M. Sergent, A.P. Ramirez, Phys. Rev. Lett. **93**, 076601 (2004)
23. O. Mitrofanov, D.V. Lang, C. Kloc, J.M. Wikberg, T. Siegrist, T.Y. So, A.M. Sergent, A.P. Ramirez, Phys. Rev. Lett. **97**, 166601 (2006)
24. M. Malagoli, J.L. Bredas, Chem. Phys. Lett. **13**, 327 (2000)
25. G.R. Hutchison, M.A. Ratner, T.J. Marks, J. Am. Chem. Soc. **127**, 2339 (2005)
26. S. Athanasopoulos, J. Kirkpatrick, D. Martinez, J.M. Frost, C.M. Foden, A.B. Walker, J. Nelson, Nano Lett. **7**, 1785 (2007)

Chapter 10
Self-Assembly of Colloidal Nanoparticles on Surfaces: Towards Surface Nanopatterning

Vasileios Koutsos, John Walker, and Emmanouil Glynos

Abstract The behaviour of colloids has become an ever expanding area of research due to the increasing number of applications in both scientific and industrial fields where their unique properties are being exploited. Such areas include bio sensors, catalyst processes, microelectronics industry and drug delivery applications. In this chapter we introduce the fundamental ideas and concepts behind the reversible self-assembly of colloidal particles on solid surfaces. The emphasis is on ultrathin films, monolayers and sub-monolayers with colloidal particles of diameter of 100 nm or lower. We provide examples of three systems (colloidal silica, magnetite and high-functionality star-shaped polymers) which highlight the importance of various interactions (electrostatic, van der Waals, steric) and small scale effects (immersion capillary forces and dewetting instabilities). Furthermore, we discuss issues associated with size and softness of the nanoparticles and the different underlying physical mechanisms that govern their behaviour.

10.1 Introduction and Theoretical Background

The world of colloidal science was once an overlooked branch of chemistry that in recent years has become a multidisciplined, billion dollar industry at the cutting edge of modern research. Applications of colloidal suspensions and their adsorption onto surfaces is the basis of many industrial processes, including waste water management, paper manufacturing, the application of paints and coatings and many chemical processes.

One of the great prospects for colloidal science is the exploitation of self-assembly techniques [1–5] for surface nanopatterning. It is a term used to describe

V. Koutsos (✉) · J. Walker · E. Glynos
Institute for Materials and Processes, School of Engineering, University of Edinburgh, King's Buildings, Edinburgh EH9 3JL, UK
e-mail: vasileios.koutsos@ed.ac.uk

S. Logothetidis (ed.), *Nanostructured Materials and Their Applications*, NanoScience and Technology, DOI 10.1007/978-3-642-22227-6_10,

the ordering of systems into structures or forms without external assistance. These structures can be formed by the initial input of some energy (static systems), by the cyclical balance of chemical reactions that are initiated and eventually dissipate (dynamic) or due to pre-patterning of the underlying substrate to facilitate a desired structure (templated) [6]. It is a facile bottom-up technique and could lead to new ways of inexpensive fabrication for a variety of industrial sectors including microelectronics, photonics and biomedical applications.

The relentless drive to shrink down the scale at which devices can be manufactured has begun to approach the physical limits of what top–down manufacturing techniques can achieve; and so attention has turned to the use of bottom-up techniques [7, 8] for manufacturing in many areas of industry. Current applications of bottom up processes using self-assembly techniques include photonics [9], biosensors [10], the semiconductor industry [1, 11] and quantum dot technology [12]. Colloidal particles have been widely used as a stock material in constructing nanostructures due to their modifiable physicochemical properties [13, 14]. This has fuelled research into the behaviour of colloidal particles in suspension [15,16], their subsequent deposition or adsorption [17–19] onto a surface and their final structuring during drying [20–28].

A *colloidal suspension* (colloid) consists of a dispersed or discontinuous phase distributed uniformly throughout a dispersion or continuous medium. Disperse systems where all the particles are of a similar dimension are known as monodisperse, and systems where the particles are present in a range of sizes are referred to as polydisperse. For the majority of colloidal materials the size of the dispersed medium lies in the dimension range of 1–1,000 nm, although this is not a set limit to define a colloidal system. This range of scale gives colloidal systems one of its defining characteristics, a larger surface area to volume ratio at the dispersed phase. Therefore it is the interfacial properties between the dispersed and continuous mediums that play a dominant role in determining the behaviour of a colloidal system.

It is normally desired that a colloidal suspension of particles remains dispersed and suspended within its medium. Due to their size colloidal particles are subject to random molecular collisions from the surrounding medium in a phenomenon known as Brownian motion. Such motion maintains the dispersion of the colloid particles throughout the colloidal suspension. Over time the effects of attractive forces that occur between particles will cause them to aggregate together until gravitation sedimentation occurs. In order to maintain a well-dispersed suspension, it is vital that the attractive forces between these particles be counterbalanced by repulsive ones, preventing particle aggregation.

The forces that drive self assembly could be of similar nature to the forces that govern the behaviour of colloidal particles in the suspension: electrostatic, van der Waals, steric and entropic forces. However, as the solvent evaporates some other important effects take place generating effective interactions, movement and new equilibrium (or in some cases non-equilibrium/kinetically driven) configurations. For example, in dried samples, that is, samples that involve the evaporation of the suspension, attractive capillary forces play a major role in the final structuring of the colloidal particles (it should be noted that capillary forces can exist within liquid systems in the form of a gas bridge between two particles). In some cases, capillary forces have been shown to be responsible for particle ordering [26] whereas electrostatic interactions facilitate particle mobility within the suspension and also deposition on the substrate, but their influence on the final structure is not as pronounced [23].

10.1.1 Colloidal Particle Interactions

The most dominant forces affecting colloidal particle behaviour until the interaction distance has reached in the region of a few nanometers (where solvation/hydration forces come into play) are van der Waals, electrostatic and steric forces. The following is a concise review on these forces that have to be considered carefully when designing self-assembly systems not only in connection with particle–particle interactions but also with particle–surface interactions. It is composed from a selection of books [29–33] and other references where noted. For a more in-depth study of the range of particle–particle and particle–surface interactions it is recommended that the referenced books are considered.

10.1.2 van der Waals Forces

The van der Waals interaction between two molecules is composed of three distinct interactions that all vary with the inverse sixth power of the separation distance. The Keesom or *orientation* interaction evaluates interactions involving permanent dipole–dipole molecules, the Debye or *induction* interaction evaluates interactions between dipole-induced dipole interactions and finally the London or *dispersion* interaction evaluates the interaction between all atoms and molecules due to the quantum induced instantaneous dipole interactions. Of the three interactions the dispersion interaction component is the most important due to it always being present (while the induction and orientation interactions are dependent on the properties of the molecules). London forces can be exhibited by nonpolar molecules because of electron density fluctuations about a molecule (based on Schrödinger equation for the variation with time of the quantum state of a physical system). When an electron is on one side of the nucleus, this side becomes slightly negative; this in turn repels electrons in neighbouring atoms, making these regions slightly positive. This *induced instantaneous dipole* causes a brief electrostatic attraction between the two molecules. The electron immediately moves to another point and the electrostatic attraction is broken. Alternatively bond vibrations in molecules may produce the oscillations or they may be triggered by random, instantaneous coalescing of electrons in atoms. The electron-rich and electron-poor regions of the induced dipole may not persist for more that 10^{-14} or 10^{-15} s, but if they can polarise the electron distribution on an adjacent molecule, electron clouds on the two molecules may begin to oscillate cooperatively with each other. The dipoles are transitory but aligned, and a net attractive force pulls the molecules together. At closer range, the oscillation becomes even more effective. The London expression for the dispersion interaction energy between two atoms or molecules is $v_L = -\frac{C_L}{r^6}$, where C_L is referred to as the London constant and depends on atomic/molecular characteristics and electromagnetic properties of the medium. Similar formulas describe the induction and orientation interactions. It is worthwhile noting that the van der Waals interactions are affected greatly by the presence of a solvent medium and also by the following: (1) The dispersion force contribution is significantly greater than that of the dipolar contribution; (2) the van der Waals interaction is significantly weakened by the presence of a solvent and (3) the dispersion force between dissimilar molecules can be attractive or repulsive. It is repulsive when the refractive index of the medium is an intermediary of the particles refractive index. For identical particles it is always attractive. Due to the fact that the period of the fluctuation in the dipoles is comparable to that of the time taken for the fluctuation to be transmitted, at long distances the dispersion energy between two atoms begins to decay even faster than $-1/r^6$, approaching $-1/r^7$ for separation distances approaching 100 nm. This phenomenon is referred to as the *retardation* effect. By assuming additivity and ignoring retardation effects, a selection of van der Waals interaction energies formulae can be derived from the integration of the interatomic van der Waals pair potential ($v_{vdW} = -C/r^6$) for sphere–sphere and sphere–surface interactions. These interaction laws are usually given in terms of the conventional Hamaker constant $A = \pi^2 C \rho_1 \rho_2$ where ρ_1 and ρ_2 are the number of atoms per unit volume in the two bodies. Typical Hamaker constants for solids and liquids in a vacuum are approximately 10^{-19} J. We note that at very short separation distances there exists a strong repulsion force generated by the overlap of the electron clouds of atoms. This force is commonly referred to as *hard core repulsion* or, for ions, the *Born repulsion*. The force is characterised by a very short range and its magnitude is rapidly increasing as the atoms approach.

10.1.3 Electrostatic Interactions

Many interfaces in an aqueous system carry an electrical charge. Interfaces with a similar charge will repel one another due to the Coulomb's law. This repulsion occurs between any "like-charge" interfaces and is an important factor in determining the colloidal behaviour of aqueous systems. Let us consider a single colloidal sphere suspended in a liquid medium. At the interface between the surface of the particle and the liquid we will assume any charged surface to be uniformly charged. Each colloid carries a "like" electrical charge which produces a force of mutual electrostatic repulsion between adjacent particles. The charging of a surface in a liquid can originate from either the ionisation or dissociation of surface groups

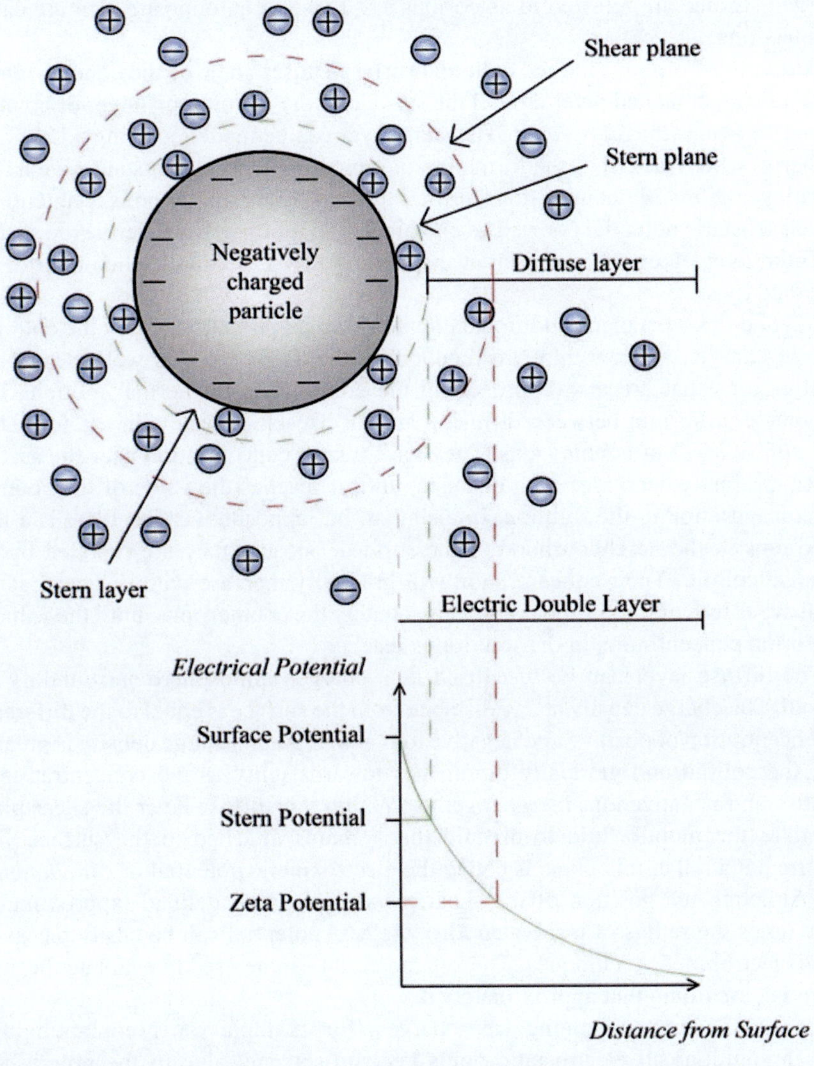

Fig. 10.1 Schematic of the Electrical Double Layer with an electrical potential graph for reference

or by the adsorption of ions from solution onto a previously uncharged surface. The final surface charge is balanced by an equal but oppositely charged region of counter-ions, some of which are bound, usually transiently, to the surface while others form an atmosphere of ions in rapid thermal motion close to the surface, known as the diffuse electrical double layer (EDL). The EDL model (Fig. 10.1) is used to visualise the ionic environment in the vicinity of a charged colloid and explains how the electrostatic repulsive force occurs. Ions of the same sign as the charged surface are referred to as co-ions and those of an opposite sign are called counter-ions.

Attraction from the charged colloidal surface causes some of the counter-ions to form a firmly attached layer around the surface of the colloid; this layer of counter-ions is known as the *Stern layer*. The stern layer can be modelled using a Langmuir isotherm, which describes the formation of a monolayer. In this instance we have a monolayer of mainly counter-ions at the surface whose population is a function of the electrostatic potential (as well as chemical interactions). If we define the surface potential as ψ_0 then the potential at the stern plane is ψ_σ due to the effect of the adsorbed ions.

Beyond the stern plane additional counter-ions are still attracted by the colloidal surface charge, but now they are repelled by the Stern layer as well as by other positive ions that are also approaching the colloid due to thermal motion. This dynamic equilibrium between diffusion and electrostatics results in the formation of a *diffuse layer* of counter-ions. They have a high concentration near the surface which gradually decreases with distance, until it reaches the value of the counter-ion concentration in the solution. In a similar, but opposite, fashion there is a lack of co-ions in the neighbourhood of the surface, because they are repelled by the negative colloid. Their concentration will gradually increase with distance, as the repulsive forces of the colloid are screened out by the counter-ions, until the value of the co-ion concentration in the solution is reached.

The diffuse layer can be visualised as a charged atmosphere surrounding the colloid. The charge density at any distance from the surface is equal to the difference in concentration of positive and negative ions at that point. Charge density is greatest near the colloid and gradually diminishes towards unity as the concentration of positive and negative ions merge together. Within this diffuse layer the shear plane separates the mobile fluid from fluid that remains attached to the surface. The electric potential at this plane is called the electrokinetic potential or *zeta potential* (ζ). Although the position of the shear plane is not well defined (approximately three times the radius of a solvated ion), the zeta potential can be easily measured using electrokinetic techniques. The zeta potential can be used to evaluate the stern potential, assuming that approximately $\zeta \approx \psi_\sigma$.

In any medium containing free charges (for example water containing free ions in solution) all electrostatic fields become screened due to the polarisation (displacement) of these charges. A screened electric field decays approximately exponentially with distance x according to $e^{-\kappa x}$ where κ is the Debye–Hückel parameter which is measured in length^{-1} (10.1). The Debye–Hückel parameter characterises the decay of the potential with the distance from the surface. The Debye screening length (κ^{-1}) is a term used to describe the characteristic length or "thickness" of the EDL.

$$\kappa = \left(\frac{\sum_i \rho_{\infty i} e^2 z_i^2}{\varepsilon_0 \varepsilon k_B T} \right)^{\frac{1}{2}} \tag{10.1}$$

where $\rho_{\infty i}$ is the ionic concentration of ions i in the bulk, e is the electronic charge constant, z_i is the valency of ion i, ε_0 is the electric constant, ε is the relative static permittivity of the medium, T is the system's absolute temperature and k_B is the Boltzmann constant. It is interesting to note that, other than some fundamental constants, the Debye length depends only on the temperature and the bulk electrolyte concentration. As such Debye lengths for known electrolyte concentrations can be quickly evaluated, for example, the Debye length of NaCl at 25°C is $1/\kappa = 0.304/\sqrt{[\mathrm{NaCl}]}$ nm where [NaCl] is the molecular concentration of the electrolyte, in this case sodium chloride. At low surface potentials (<25 mV) the potential of the EDL becomes proportional to the surface change density, and can be calculated as a function of the distance away from the stern layer as $\psi(x) \approx \psi_\sigma e^{-\kappa x}$.

10.1.4 DLVO Theory

Colloidal particle–particle interactions are described using the Derjaguin–Landau–Verwey–Overbeek (DLVO) [34] theory. The DLVO model combines the basic forces governing colloidal particles in suspension and it is primarily based on the relationships of long-range repulsive electrostatic forces and short-range attractive forces from the van der Waals interactions.

The stability of a particle in solution is dependent upon its total potential energy function V_T. DLVO theory assumes that V_T is the balance of two competing contributions:

$$V_T = V_A + V_R \tag{10.2}$$

where V_A is the non-retarded van der Waals attractive potential and V_R the EDL repulsive potential. The attractive potential energy related to van der Waals force interactions as the separation distance, D, between spherical particles of radius R changes is given by

$$V_A = -\frac{AR}{12D} \tag{10.3}$$

The repulsive potential V_R becomes significant when two colloids approach each other and their double layers begin to interfere. It has a maximum value when they are almost touching and decreases to zero outside the EDL.

$$V_R = \left(\frac{64\pi k_B T R \rho_\infty \gamma^2}{\kappa^2}\right) e^{-\kappa D} \tag{10.4}$$

where ρ_∞ is the electrolyte concentration in the bulk, $\gamma = \tanh(ze\,\psi_0/4k_B T)$.

DLVO theory allows evaluations of the stability of a colloidal system to be determined by the sum of these attractive and repulsive forces that exist between particles as they interact with each other. A system could have a primary minimum and primary maximum in the total interaction energy potential. The primary minimum is where the attractive interactions dominate over the repulsive interactions and, conversely, the primary maximum is where the repulsive interactions dominate over the attractive interactions. In order to reach the primary minimum the particles must exceed the primary maximum activation energy. Therefore if $V_{\mathrm{T}}^{\mathrm{MAX}} \gg k_{\mathrm{B}}T$ the particles will be in a stable colloidal state. In certain situations (e.g. high electrolyte concentrations decreases the Debye length, effectively reducing the extent and intensity of the repulsive electrostatic interactions between the colloidal particles), there is a possibility of a secondary minimum where a much weaker and potentially reversible aggregation between particles exists. At the secondary minimum $-V_{\mathrm{T}}^{\mathrm{SMIN}} \sim k_{\mathrm{B}}T$ so the net attractive energy is at best only slightly larger than the average thermal energy. The result of this is the formation of weak flocs, sufficiently stable not to be broken up by Brownian motion, but may dissociate under an externally applied force such as vigorous agitation.

More recently the standard DLVO model has been updated to incorporate the short-range Lewis acid–base (AB) interactions that account for electron acceptor/electron donor interactions which have a measurable effect within a few nm of separation distance. The extended DLVO or XDLVO theory also can encompass other non-DLVO interactions arising from solvation forces in aqueous systems, in particular hydration forces [35] due to the energy required to disrupt the hydrogen-bonding network formed by the binding of water molecules to hydrophilic surfaces and structuring. The hydration force is an oscillating repulsive force of periodicity roughly equal to the diameter of a water molecule that grows exponentially in magnitude. In the case of silica and mica it is believed to arise from strongly H-bonding surface groups such as hydrated ions or hydroxyl groups, which leads to the modification of the H-bonding network of liquid water adjacent to them.

10.1.5 Electrolyte Concentration Control over Interactions

Due to the dependence of the Debye length on the ion concentration of the suspension, modification of the electrolyte concentration can provide us with a mechanism by which we can directly control the colloidal particle interactions. As attractive van der Waals forces are dependent only on the separation of the particles/surfaces, at low electrolyte concentrations when Debye screening is minimal, long-range repulsions severely limit the extent of adsorption due to the extended double-layer of ions surrounding it [36]. At higher electrolyte concentrations, the reduction in the Debye length causes the distance over which effective interparticle and particle–surface repulsions occur to be reduced. This suppression of the effective range of the repulsive force allows particles to approach one another and other surfaces at closer ranges, facilitating more densely packed arrangements, thus attaining higher surface coverage per unit area [37].

10.1.6 Steric Interactions

As previously mentioned, in order to maintain the stable dispersed colloidal system, the repulsive forces must be dominant ($V_{\mathrm{T}}^{\mathrm{MAX}} \gg k_{\mathrm{B}}T$). To achieve an effective repulsive force without the use of electrostatic potentials, steric repulsion potentials can be utilised. For colloidal stability the use of steric repulsion involves the application of a thermally diffuse interface of chain molecules (oligmer or polymer) attached at some point to the surface of the colloid dangling out into the solution where they are thermally mobile. On approach of an analogous colloidal particle the entropy of confining these dangling chains results in a repulsive entropic force referred to as the "steric" or "overlap" repulsion. This allows colloidal particles suspended in a non-polar medium to be stabilised against coagulation by the addition of a suitable polymer to the system adsorbing onto the colloidal particle surface and physically preventing the particle surfaces coming into close enough contact for attractive London–van der Waals forces to cause coagulation. Unlike electrostatic stabilisation, there are no long-range repulsive forces and the particles could be subject to attractive forces until the outer portions of the steric molecules contact each other. The magnitude of the repulsion depends on the molecular weight of the polymer chain, the coverage of the polymer on the surface of the colloid, the mechanism by which the polymer is attached to the surface (adsorbed or end-grafted) and the quality of the solvent (i.e. if the polymer is in a poor solvent will coil up and shrink).

10.2 Experimental

10.2.1 Atomic Force Microscopy

The atomic force microscope was first reported by Binnig, Quate and Gerber [38] of IBM in 1986 and followed on from their work on scanning tunnelling microscopy (STM) [39], for which they were awarded the Nobel Prize for Physics. It was suggested as a means of studying non-conducting surfaces on an atomic scale, combining the principles of the STM and a stylus profilometer, and imaging by effectively "feeling" the sample surface. Owing to its high resolution (sub-nm) and ease of use in ambient conditions or within liquids, it is the most suitable technique for the direct observation of nanoparticle assemblies on surfaces and it is used widely.

The major components of a modern AFM are given in Fig. 10.2. The most important part is the tip, which makes the physical contact with the surface of the sample. The tip is usually made of silicon or silicon nitride, which is harder and hence more wear resistant. Typically tips have a micrometer scale pyramidal shape with a nanometer size apex radius. The tip is connected to a cantilever which allows the tip to move in relation to the topography of the sample. A laser beam

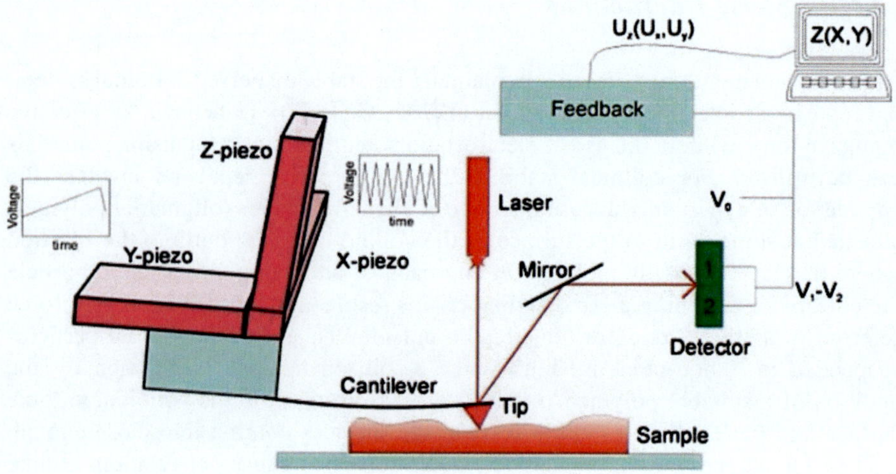

Fig. 10.2 Schematic diagram of an atomic force microscope

is directed onto the backside of the cantilever at the tip end and its reflection is directed onto a photodiode detector with four quadrants. The intensity of light in each quadrant gives information on the position of the cantilever and hence the tip. This information is then sent to a controller where a feedback mechanism registers a deflection of the cantilever and manipulates the displacement between the tip and the sample so that the deflection is constant. This is achieved by using piezoelectric crystals which can be utilised either as the sample stage or, in the case of our schematic, as the cantilever mount. When a voltage is applied across it, a controlled expansion/contraction occurs in the piezoelectric crystal. This motion is very reproducible and precise and allows the crystal to be deformed with the accuracy of atomic dimensions, giving an AFM the ability to perform very precise measurements of the sample topography. An image of the sample is achieved by raster scanning i.e. scanning line by line in the X and Y directions (again by using piezoelectric crystals to control the movement). The Z height data is calculated by taking the varying voltage applied to the Z axis piezoelectric crystal to maintain a constant deflection and scaling it with the known calibrated ratio of voltage to distance of the piezoelectric crystal.

This mode of operation is referred to as contact mode AFM: the cantilever/tip is maintained at a constant deflection and moved along the surface. This is one of the simplest methods of imaging using AFM techniques, however it has several drawbacks. The tip can exert considerable force on the sample surface. For soft samples such as polymers or biological specimens, the tip-force may cause an irreversible deformation of the surface so that the topography information becomes ambiguous or inaccurate. Similarly the frictional and lateral forces generated by the dragging of the tip also have a large magnitude. As such any weakly adsorbed material on the substrate will simply be dragged along the surface, rendering contact mode ineffective for the imaging of delicate samples. To overcome such difficulties, it was suggested in 1987 that a non-contact (NC) mode [40] could be employed whereby the tip is not in constant contact with the surface but rather vibrates near the surface at its resonance frequency.

Intermittent contact (IC) mode AFM, was introduced in 1993 and differs only slightly from NC mode in that at the extreme of each oscillation the tip touches the surface very briefly. As the tip is brought close to the sample surface in IC mode, the characteristics of the cantilever vibration (e.g. amplitude, resonance frequency) change due to the tip-sample interaction and the feedback mechanism adjusts the tip-sample separation in order to maintain a constant amplitude of vibration. Another development of IC mode AFM gives the ability to detect shifts in phase angles of vibration when the oscillating cantilever interacts with the sample surface. The detection of phase angle shifts provides enhanced image contrasts for heterogeneous surfaces.

Given the size of the particles to be imaged in the present study, one of the most important AFM artefact effects to consider is tip convolution. This occurs due to the shape and the finite sharpness of the tip. The conical or pyramid or possibly spherical shape (at its very end) of a tip means that the side of the tip will make physical contact before the tip end, resulting in the imaged particle of radius R appearing wider than it really is, while the apparent height, $2R$, will be accurately measured as long as the particles are relatively sparsely distributed. Thus, it is important to accurately match the tip size to the topography of the sample. While we expect that some degree of blunting will occur throughout our experiments we endeavour to use clean, sharp tips for each set of experiments in order to avoid excessive convolution of the images that can be avoided if possible.

The most widely used substrate for AFM experiments is mica. Mica is a group of sheet silicate (phyllosilicate) minerals that feature highly perfect basal cleavage, due to the hexagonal sheet-like arrangement of its atoms. This property provides us with an easy to prepare, atomically flat substrate. For our experiments we will be using muscovite, $KAl_2Si_3AlO_{10}(OH, F)_2$, a high-aluminium mica. Upon cleavage along the basal plane, the exposed surface of the mica is a highly hydrophilic, negatively charged surface due to the non-equilibrium distribution of potassium ions across the cleaved surface, which reduces in time due to the attraction of oppositely charged contaminations.

10.3 Drying and Immersion Capillary Forces: The Emergence of Order

Lateral capillary forces between colloids in thin films occur when the particles are partially immersed in the liquid layer; this immersion capillary force causes the liquid surface surrounding the particle to deform as the liquid wets the surface of the colloidal particle. As a liquid film evaporates its height from the substrate shrinks until any colloidal particles within the film protrude through it. When this happens,

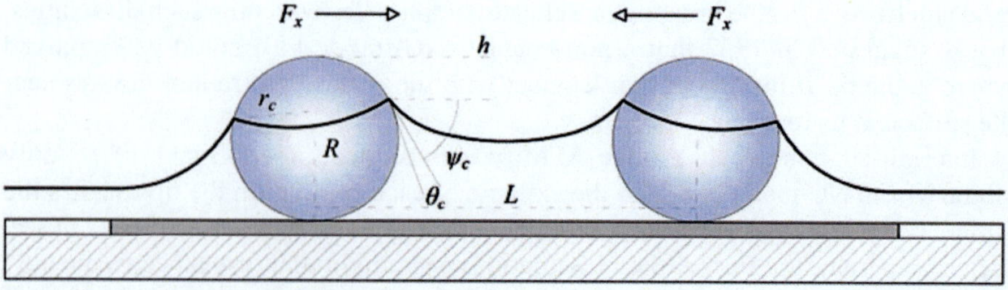

Fig. 10.3 Two spheres partially immersed in a liquid layer on a horizontal solid substrate. The deformation of the liquid meniscus and the particle separation distance gives rise to interparticle attraction

menisci form around the tops of the colloidal particles, generating an attractive force due to the deformation of the liquid surface and the induced asymmetry of the contact line at the surface of the particle. The magnitude of the force experienced between two neighbouring particles is dependent on several aspects including the separation distance of the particles, the radius of the three phase contact line and the contact angle of the meniscus. The latter two are dependent on the particle size and liquid film height.

Figure 10.3 gives a schematic representation of two spheres of radius R on a surface under the influence of capillary force F_x which can be approximated [27] as:

$$F_x \approx 2\pi\sigma r_c^2(\sin^2\psi_c)(1/L) \tag{10.5}$$

where σ is the surface tension of the liquid, r_c is the radius of the three-phase contact line at the particle surface ($r_c = [h(2R-h)]^{1/2}$), ψ_c is the mean meniscus slope angle at the contact line ($\psi_c = \arcsin(r_c/R)$–θ_c, where θ_c is the contact angle of the bulk liquid), h is the height of the liquid layer from the top of the particle ($2R$), and L is the distance between the particles. It is clear from the equation that the radius of the particle R plays a dominant role in dictating the capillary force experienced between the colloidal particles. Also notice that force exerted by the menisci is proportional to the inverse of the distance between the neighbouring particles (L). This means that the further the neighbouring particles are apart when the menisci forms between them, the less force the capillary action exerts between them.

These capillary forces are the main driving force behind creating self assembled 2D colloidal crystals. These are self assembled regular patterns that may extend to areas of mm^2 and are of great interest to the photonics industry due to their Bragg diffraction properties [24]. Early techniques used for constructing 2D crystal films were simply droplets of the colloidal suspension evaporated onto a substrate. Refinement of this process came with the utilisation of the meniscus line in evaporating liquids and by the 1990s the mechanism and governing forces of the 2D particle assembly had been clarified [26]. A two stage process was observed: (1) nucleus formation, under the action of attractive capillary immersion forces; and (2) crystal growth, through convective particle flux caused by the water evaporation from the already ordered array [24]. Control over this capillary force comes from

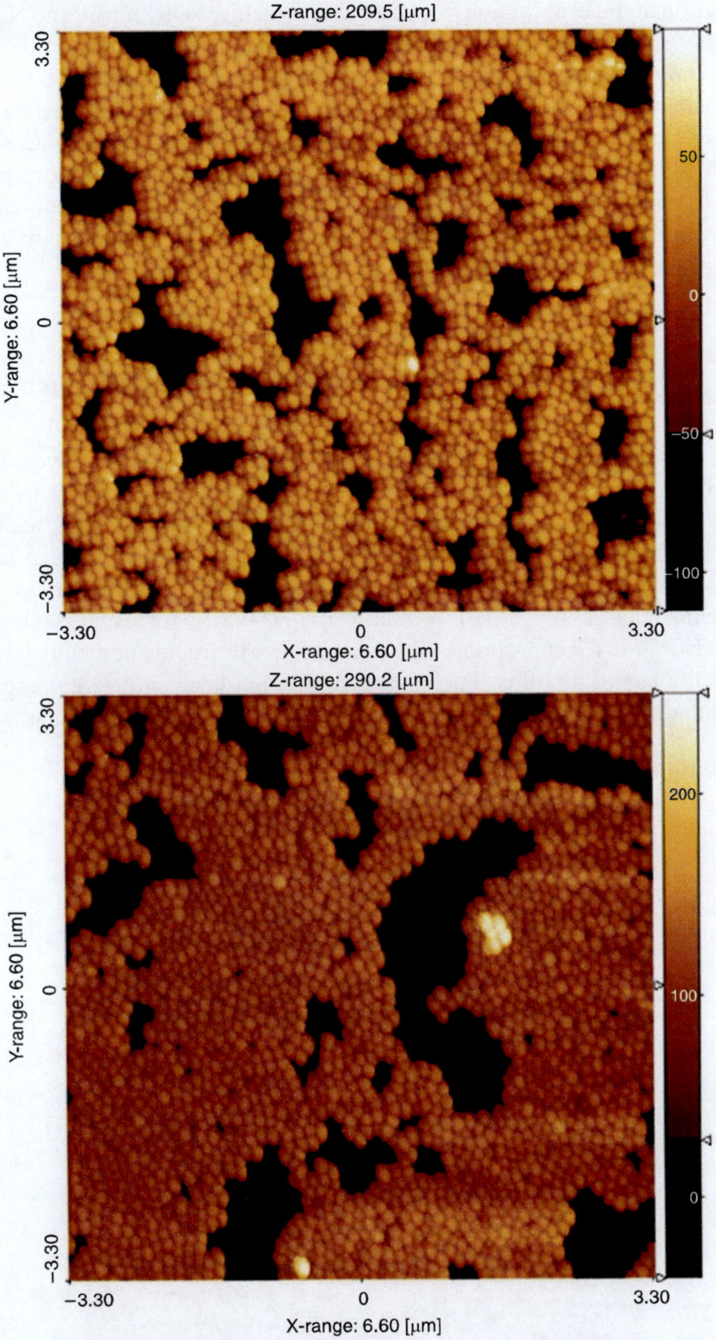

Fig. 10.4 $6.6 \times 6.6\,\mu m^2$ AFM images from different samples of 100 nm silica particles, diluted to 10% by volume. Freshly cleaved mica was submerged in the suspension and dried with a nitrogen stream

manipulation of the evaporation rate and the film meniscus. Dimitrov and Nagayama [26] demonstrated a now well-established technique for creating tuneable 2D crystalline films using a setup akin to a dip coating device, where the substrate is submerged into the colloidal suspension then withdrawn at a controlled rate. Other techniques include droplet application followed by tilting of the substrate [41], spin coating the colloidal suspension on the substrate [42] and the use of pre-patterned surfaces to create specific sites that are more energetically favourable to adsorb to. These techniques result in 2D crystalline films where either multidomain or one large single domain crystal is grown.

10.3.1 Crystalline Monolayers of Colloidal Silica on Mica

When suspended in water, silica particles [43] have a native negative charge on the surface due to the dissociation of silanol groups which allow them to maintain an even dispersion in favourable pH levels and avoid coagulation without the need for surface chemistry modification. The system of silica particles and mica gives us an opportunity to study self-assembly of nanoparticles on a substrate of the same charge (negative). While van der Waals forces are sufficient to provide adsorption for appropriate incubations times the repulsive electrostatic force enhances the lateral mobility. Such systems have not been sufficiently explored in current literature. They provide an alternative system for bottom up self assembly of

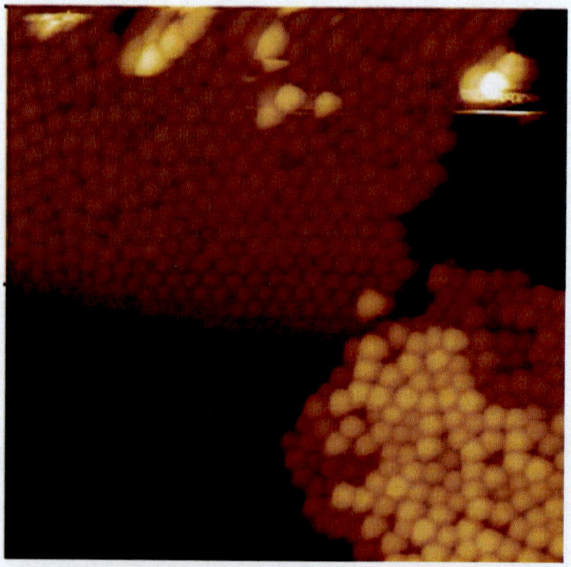

Fig. 10.5 3.3 × 3.3 μm^2 topography AFM scan with a colloidal crystalline monolayer area at the upper part of the image

colloidal nanoparticles. In Fig. 10.4 we show silica nanoparticles (Polysciences Inc., Baden-Württemberg, Germany) of diameter of ~100 nm deposited on mica by dip coating and subsequent drying. The monolayer network observed is characteristic of the drying and dewetting process while the ordering (see also zoom in Fig. 10.5) is attained by crystal nucleus formation due to attractive capillary immersion forces and subsequent crystal growth via convective particle flux as the water evaporates. The repulsive electrostatic interactions between silica and mica (as they are both negatively charged) facilitate mobility along the surface.

10.4 Dewetting Effects: Self-Organisation

The crystalline structuring of colloidal particles has typically been the domain of some hundreds of nanometers to micrometer sized particles. Research into the self assembly of nanometer sized colloids has yielded some radically different particle structuring, such as 2D cellular structures. These are normally associated with dewetting phenomena found in thin liquid films that undergo rupture and coalescence into network structures. Martin et al. [44] carried out studies linking this, normally fluid, behaviour to other systems including nanocolloidal suspensions of gold nanoparticles of a few nanometers in diameter. They found striking resemblances in morphology and using Voronoï tessellation to calculate the distribution of particles on the surface of the samples were able to assess the degree of (dis)order within a system by comparing these results with those from Poisson distributed simulations.

Rabani et al. [45] demonstrated a model for drying mediated self assembly of nanoparticles that simulates remarkably well the behaviour of these complex fluids by including not only the relatively weak attractions between the particles themselves but also the dynamics of an evaporating solvent. Using their model they were able to show that varying the choice of solvent, particle size and thermodynamic state gave rise to various morphologies.

The dewetting phenomena can simply be described with the example of a new car left in the rain. The water droplets on the waxed surface do not coat the bodywork evenly, rather they coalescence into droplets on the surface. In effect the rain water "dewets" on the car panels. Generally dewetting describes the rupture of a thin liquid film on the substrate (either a liquid itself, or a solid) and the subsequent formation of droplets. Recent insights into the balance between short and long-range forces involved in dewetting phenomena resulted in Sharma and Reiter [46, 47] presenting the following four-stage dewetting process from experimental observations: (1) rupture of the thin film; (2) expansion and coalescence of holes to form polygonal "cellular" patterns; (3) fingering instability of hole rims during hole expansion witnessed only on low wettability surfaces (can also occur after stage 4); (4) disintegration of liquid ridges forming the polygon into spherical drops due to Rayleigh instability.

Seemann, Herminghaus and Jacobs [48] provided a clarification of the distinction between stable, unstable and metastable films using the effective interface potential, $\Phi(h)$, which is defined as the excess free energy (per unit area) it takes to bring two interfaces from infinity to a certain distance, h. In the case of a dewetting film, the two interfaces involved are the solid/liquid interface and the liquid/air interface, and h is the initial thickness of the liquid film. By consequence $\Phi \rightarrow 0$ as $h \rightarrow \infty$. In a scenario where $\Phi(h) > 0$ the global minimum lies at infinite film thickness. In this case the liquid film is termed stable. If the global minimum of $\Phi(h)$ occurring at a point h^* then the free energy $\Phi(h)$ of the film possesses a negative curvature at the starting thickness, h_o, i.e. $\Phi''(h_o) < 0$, the system can gain energy by allowing the film thickness to reach h^*. If the initial film thickness is greater than h^* then the film will try to attain an equilibrium film thickness of h^* causing localised thinning in the film that will ultimately lead to rupturing and subsequent dewetting. Thus from infinity to h^* the film is unstable against *spinodal* dewetting. Spinodal Dewetting [49] is triggered by spontaneous amplification of capillary waves within the film caused by thermal fluctuations and is usually characterised by a bicontinuous structure of the phase separation morphology which can be identified by 2D-FFT analysis of the morphology. Finally, there is a third scenario where a system is metastable; that is at low film thicknesses where $\Phi''(h_o) < 0$, spinodal dewetting can occur, but at larger film thicknesses it is stable against spinodal dewetting.

A second dewetting mechanism known as nucleation dewetting [50] can cause dewetting to occur on stable, metastable and unstable films either by nuclei defects such as dust particles/surface heterogeneities (heterogeneous) or by localised thermal instabilities (homogeneous). Nucleation dewetting can occur in parallel or at a different time frame to spinodal dewetting as the two processes are independent.

For heterogeneous nucleation, because the dewetting is initiated by surface defects the initial surface ruptures occur over a small time frame. Unlike spinodal dewetting however a characteristic length in the morphology does not exist. Homogeneous nucleation differs in that it has a continuous breakout of holes on the surface throughout the time frame, caused by the local fluctuations in thermal energy allowing the liquid to overcome the potential barrier for nucleation of a dry spot, leading to the formation of a hole. There are usually time and size differences between the different dewetting mechanisms.

10.4.1 Dewetting Structures of Colloidal Magnetite Nanoparticles on Mica

In Fig. 10.6, we present typical images of colloidal deposits after evaporation of a colloidal magnetite n-heptane suspension [51]. In this case the nanoparticles have been much smaller: diameter ~10 nm. The solvent film ruptured at a higher thickness and before immersion capillary forces were able to be exerted on the individual nanoparticles. The end result was the nanoparticles to follow the solvent

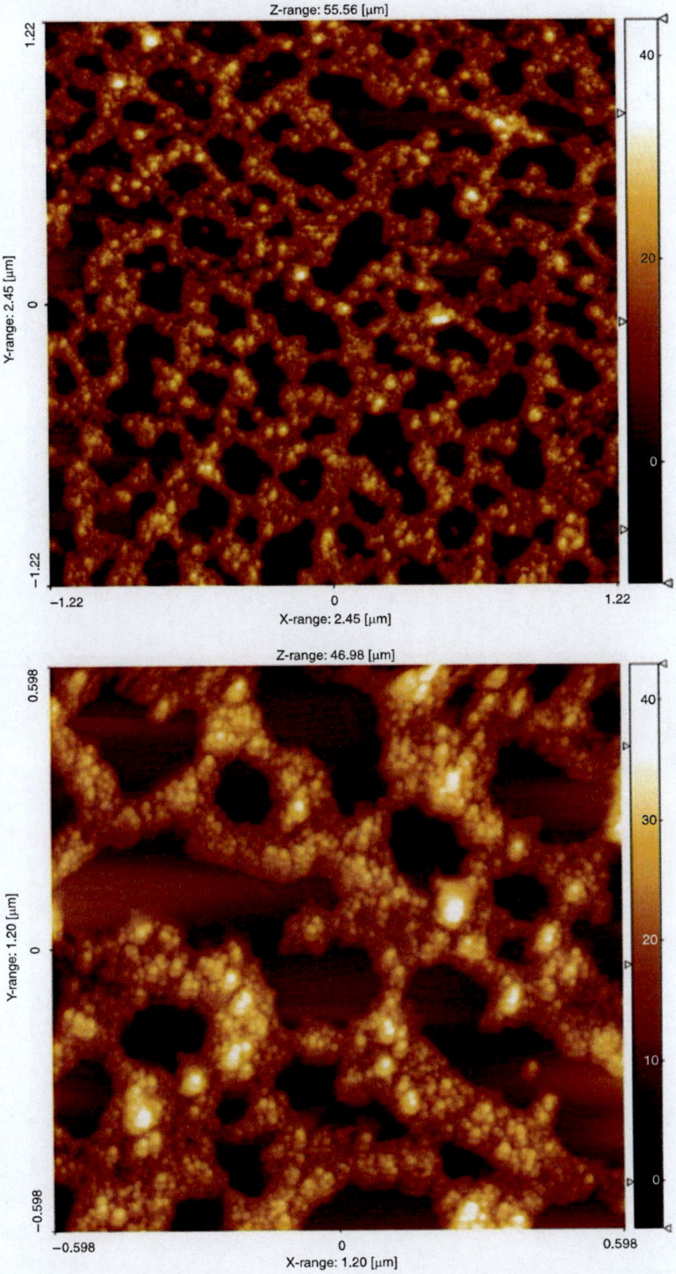

Fig. 10.6 AFM topography images (top 2.45 × 2.45 μm^2, bottom 1.20 × 1.20 μm^2) of deposits on mica after solvent evaporation from dilute magnetite nanoparticle suspensions

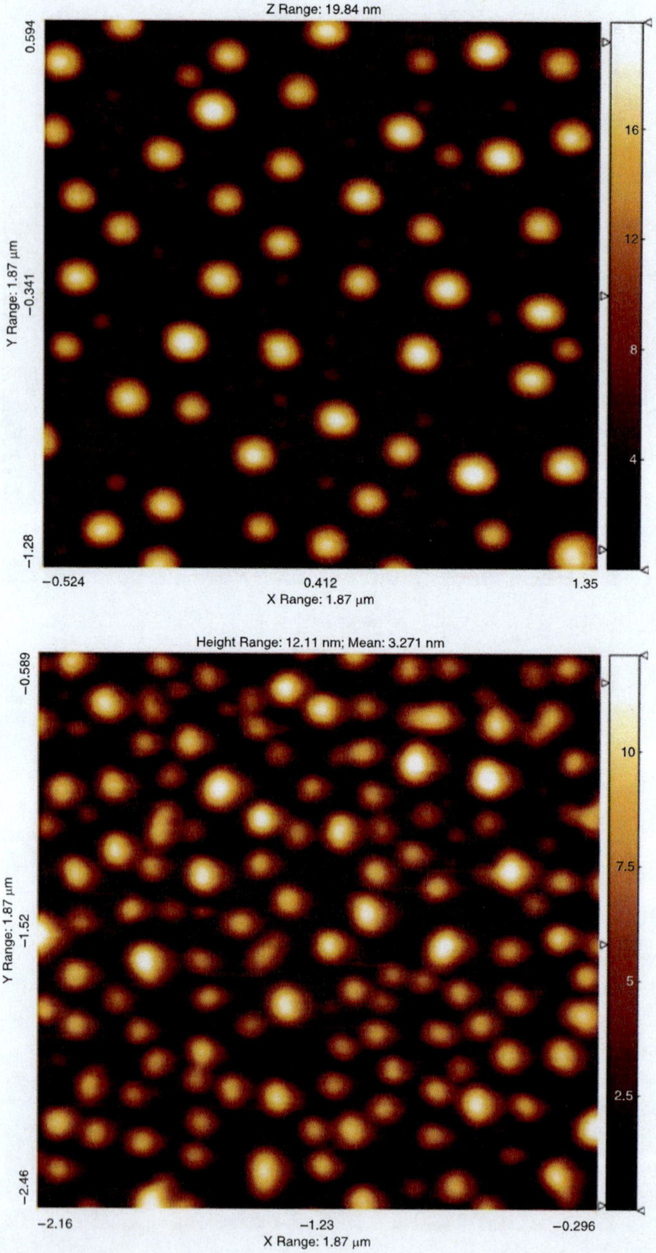

Fig. 10.7 AFM topography images ($1.87 \times 1.87\,\mu m^2$) of 59-arm star polymers adsorbed on mica. Top: after 1,080 min, bottom: after 3,600 min immersion time in the toluene solution

dewetting instability en mass and to create an irregular (non-crystalline) multilayer network (see zoom in Fig. 10.6). It is worthwhile noting the profound difference in the fine structure of 100 nm particles (Fig. 10.5) and 10 nm particles (Fig. 10.6). The 'premature' rupture of the film does not allow the immersion capillary forces to develop and create order in the case of the smaller particles. Nevertheless, there is still the cellular network formation which is still evidence for self-organisation resulting from the dewetting process.

10.4.2 Adsorption and Self-Assembly of Soft Colloid Nanoparticles on Mica

Figure 10.7 presents AFM images of a soft nanoparticle system: mica substrates dip-coated in a 59-arm polybutadiene star polymer toluene solution [52, 53]. Star polymers (macromolecules with homopolymer arms covalently joined to a dendritic core) with a large number arms (high functionality) behave like colloid particles owing to the fact that the osmotic pressure within the star increases with the star polymer functionality, which in turn makes the star polymer harder, preventing interpenetrations between different star polymers [54, 55]. The star polymers of such functionality behave to some extent like colloid particles stabilised by steric repulsions that develop within a good solvent (toluene in this case). In solution but also in dry state they keep to some extent their 3D shape [52, 53], they do not fuse and keep their individuality even at high surface densities (Fig. 10.7). However, it is quire remarkable that they do not seem to be affected by capillary forces and dewetting instabilities even though their interactions with the surface are relatively weak at the monomer level (just dispersion forces as polybutadiene is apolar). These polymer particles possess surface mobility but the final position of the particles are determined solely by particle–particle interactions on the surface when in the solution (mainly steric repulsions in this case); the AFM images are essentially snapshots of the polymer particle position just before drying [52, 53]. This can be an advantage for self assembly allowing more complex interactions (than capillary forces) to take place and thus more complex structures to develop on the surface (while in solution) without having the complication of solvent instabilities during drying.

10.5 Conclusions

We have shown that colloidal nanoparticles can be used for nanopatterning of solid surfaces via the self-assembly route. This process provides an inexpensive and facile methodology for the fabrication of large areas of nanoscale patterns. However, the mechanisms that govern the self-assembly behaviour are complex and in many cases competing. We have provided examples to demonstrate these factors in model systems. While relatively large nanoparticles (~100 nm) can form colloidal crystalline monolayers via the immersion capillary forces, the situation becomes very different at diameters of 10 nm, where dewetting effects dominate. However, neither of these mechanisms seem to dominate the behaviour of soft colloidal nanoparticles (high functionality star polymers) which self-assemble based purely on molecular interactions (steric repulsions) within the solution even when they are weakly adsorbed on the substrate (via van der Waals forces). Such systems could offer opportunities for the finetuning of interactions to fabricate more complex nanopatterns.

References

1. J.H. Fendler, Chem. Mater. **13** 3196–3210 (2001)
2. J. Texter, M. Tirrell, AIChE J. **47** 1706–1710 (2001)
3. G.M. Whitesides, B. Grzybowski, Science **295** 2418–2421 (2002)
4. S.C. Glotzer, M.J. Solomon, N.A. Kotov, AIChE J. **50** 2978–2985 (2004)
5. C.P. Martin, M.O. Blunt, E. Vaujour, A. Fahmi, A. D'Aléo, L. De Cola, F. Vögtle, P. Moriarty, Chapter 1, Self-Organised Nanoparticle Assemblies: A Panoply of Patterns, In: N. Krasnogor, S. Gustafson, D.A. Pelta and J.L. Verdegay, Editors, Studies in Multidisciplinarity, Elsevier, Oxford, 2008, Volume 5, Systems Self Assembly: Multidisciplinary Snapshots, Pages 1–20
6. G.M. Whitesides, B. Grzybowski, Science **295** 2418–2421 (2002)
7. M. Shimomura, T. Sawadaishi, Curr. Opin. Colloid Interface Sci. **6** 11–16 (2001)
8. D. Mijatovic, J.C.T. Eijkel, A. van den Berg, Lab Chip **5** 492–500 (2005)
9. S.H. Im, Y.T. Lim, D.J. Suh, O.O. Park, Adv. Mater. **14** 1367–1369 (2002)
10. O.D. Velev, E.W. Kaler, Langmuir **15** 3693–3698 (1999)
11. P. Hanarp, D.S. Sutherland, J. Gold, B. Kasemo, Colloids Surf. A Physicochem. Eng. Asp. **214** 23–36 (2003)
12. P. Guyot-Sionnest, Compt. Rend. Phys. **9** 777–787 (2008)
13. M. Qhobosheane, S. Santra, P. Zhang, W.H. Tan, Analyst **126** 1274–1278 (2001)
14. M.S. Romero-Cano, A. Martin-Rodriguez, G. Chauveteau, F.J. de las Nieves, J. Colloid Interface Sci. **198** 266–272 (1998)
15. J.Z. Xue, E. Herbolzheimer, M.A. Rutgers, W.B. Russel, P.M. Chaikin, Phys. Rev. Lett. **69** 1715–1718 (1992)
16. M. Elimelech, C.R. Omelia, Langmuir **6** 1153–1163 (1990)
17. Z. Adamczyk, J. Colloid Interface Sci. **229** 477–489 (2000)
18. Z. Adamczyk, L. Szyk, Langmuir **16** 5730–5737 (2000)
19. Z. Adamczyk, B. Siwek, M. Zembala, Colloids Surf. A Physicochem. Eng. Asp. **62** 119–130 (1992)
20. N. Rana, S.T. Yau, Nanotechnology **15** 275–278 (2004)
21. S. Maenosono, T. Okubo, Y. Yamaguchi, J. Nanopart. Res. **5** 5–15 (2003)
22. N. Samid-Merzel, S.G. Lipson, D.S. Tannhauser, Physica A **257** 413–418 (1998)
23. J. Aizenberg, P.V. Braun, P. Wiltzius, Phys. Rev. Lett. **84** 2997–3000 (2000)
24. P.A. Kralchevsky, N.D. Denkov, Curr. Opin. Colloid Interface Sci. **6** 383–401 (2001)
25. P.A. Kralchevsky, K. Nagayama, Langmuir **10** 23–36 (2002)
26. A.S. Dimitrov, K. Nagayama, Langmuir **12** 1303–1311 (1996)
27. N.D. Denkov, O.D. Velev, P.A. Kralchevsky, I.B. Ivanov, H. Yoshimura, K. Nagayama, Langmuir **8** 3183–3190 (1992)
28. K. Nagayama, Colloids Surf. A Physicochem. Eng. Asp. **109** 363–374 (1996)

29. Z. Adamczyk, *Particles at Interfaces, Volume 9: Interactions, Deposition, Structure*, (Academic Press, Amsterdam, Oxford, 2006)
30. D.H. Everett, *Basic Principles of Colloid Science* (Royal Society of Chemistry, London, 1988)
31. J. Goodwin, *Colloids and Interfaces with Surfactants and Polymers: An Introduction* (John Wiley, Chichester, 2004)
32. R.J. Hunter, *Foundations of Colloid Science* (Oxford University Press, Oxford, 2001)
33. J.N. Israelachvili, *Intermolecular and Surface Forces* (Academic Press, New York, 1992)
34. B. Derjaguin, L.D. Landau, Acta Physicochim. (USSR) **14** 633–662 (1941)
35. R.M. Pashley, J. Colloid Interface Sci. **80** 153–162 (1981)
36. R. Rajagopalan, R.Q. Chu, J. Colloid Interface Sci. **86** 299–317 (1982)
37. C.A. Johnson, A.M. Lenhoff, J. Colloid Interface Sci. **179** 587–599 (1996)
38. G. Binnig, C.F. Quate, C. Gerber, Phys. Rev. Lett. **56** 930 (1986)
39. G. Binnig, H. Rohrer, C. Gerber, E. Weibel, Phys. Rev. Lett. **50** 120 (1983)
40. Y. Martin, C.C. Williams, H.K. Wickramasinghe, J. Appl. Phys. **61** 4723–4729 (1987)
41. R. Micheletto, H. Fukuda, M. Ohtsu, Langmuir **11** 3333–3336 (1995)
42. P. Jiang, T. Prasad, M.J. McFarland, V.L. Colvin, Appl. Phys. Lett. **89**011908–011903 (2006)
43. R.K. Iler, *The Chemistry of Silica: Solubility, Polymerization, Colloid and Surface Properties, and Biochemistry* (Wiley, New York, 1979), pp. xxiv, 866
44. C.P. Martin, M.O. Blunt, P. Moriarty, Nano Lett. **4** 2389–2392 (2004)
45. E. Rabani, D.R. Reichman, P.L. Geissler, L.E. Brus, Nature **426** 271–274 (2003)
46. A. Sharma, G. Reiter, J. Colloid Interface Sci. **178** 383–399 (1996)
47. A. Sharma, R. Verma, Langmuir **20** 10337–10345 (2004)
48. R. Seemann, S. Herminghaus, K. Jacobs, Phys. Rev. Lett. **86** 5534–5537 (2001)
49. R. Xie, A. Karim, J.F. Douglas, C.C. Han, R.A. Weiss, Phys. Rev. Lett. **81** 1251–1254 (1998)
50. U. Thiele, M.G. Velarde, K. Neuffer, Phys. Rev. Lett. **87** 016104 (2001)
51. K.J. Mutch, V. Koutsos, P.J. Camp, Langmuir **22** 5611–5616 (2006)
52. E. Glynos, A. Chremos, G. Petekidis, P.J. Camp, V. Koutsos, Macromolecules **40** 6947–6958 (2007)
53. A. Chremos, E. Glynos, P.J. Camp, V. Koutsos, Soft Matter **6** 1483–1493 (2010)
54. D. Vlassopoulos, G. Fytas, T. Pakula, J. Roovers, J. Phys.: Condens. Matter **13** R855–R876 (2001)
55. C.N. Likos, Phys. Rep. **348** 267–439 (2001)

Index

S. Logothetidis (ed.), *Nanostructured Materials and Their Applications*, NanoScience and Technology, DOI 10.1007/978-3-642-22227-6,